The Telecommunications Emergency Response Guide

MARCOS DUARTE RIVERO

The Telecommunications Emergency Response Guide

Restoring Critical Communications in Disaster Zones

Part of the BUSINESS FIRST AID Series

Marcos Duarte Rivero

The Telecommunications Emergency Response Guide

First Edition: March 2025

BUSINESS FIRST AID
"Saving companies"

https://www.businessfirstaid.online

Disclosure

This guide adapts emergency response principles to the telecommunications context as an instructional framework only. I want to be clear that I do not equate the importance of restoring communications infrastructure with the sacred work of saving human lives. The knowledge and skills of emergency medical professionals represent the highest form of expertise and service to humanity.

My deepest respect and profound gratitude extend to emergency responders worldwide who risk their own safety and well-being to save others. The principles they follow and the protocols they implement are the result of rigorous training and countless hours of practical experience dealing with life-or-death situations.

This guide is intended solely as a practical, accessible tool for telecommunications professionals who need a systematic approach for responding to communications failures during disasters. By borrowing the structure and approach of emergency response, I hope to provide a clear pathway for action during the chaos and pressure of disaster situations.

My aim is not to diminish or trivialize the profound work of emergency professionals but rather to honor their methodologies by showing how their systematic approach can inform other critical situations, albeit ones with different stakes. If this guide helps restore vital communications that enable emergency coordination, reconnect isolated communities, and support recovery efforts, it will have served its purpose while paying humble tribute to the emergency response principles that inspired it.

INTRODUCTION: WHEN COMMUNICATIONS FALL SILENT

The deafening roar of Hurricane Otis's 165 mph winds had finally subsided, but the silence that followed was even more alarming. Acapulco and the surrounding Guerrero region, once humming with the constant digital pulse of modern telecommunications, had gone dark. No cell service. No internet. No way for emergency responders to coordinate, for families to check on loved ones, or for authorities to assess the devastation. In that moment of digital silence, the critical role of telecommunications in disaster response became starkly apparent.

As our emergency telecom response team navigated the debris-strewn roads toward Acapulco in October 2023, we couldn't yet grasp the full extent of the destruction. Hurricane Otis had intensified from a tropical storm to a Category 5 hurricane in less than 24 hours, leaving no time for normal preparation protocols. Cell towers were toppled, fiber lines severed, power infrastructure decimated. The communications networks that modern emergency response depends upon had vanished precisely when they were needed most.

What happened in the following 48 hours would determine whether emergency responders could coordinate effectively, whether trapped victims could be located, and whether aid could be distributed where it was most desperately needed. The clock was ticking—not on an abstract deadline, but on the very real survival chances of people waiting for rescue.

The Critical Role of Telecommunications in Disaster Response

When a major disaster strikes, the restoration of telecommunications services becomes an urgent priority that enables all other emergency response efforts. These services provide the foundation for:

Understanding the disaster's true magnitude. Without reliable communications channels, the outside world cannot accurately assess the scale of destruction or the scope of humanitarian needs. During Hurricane Otis, the first satellite connection we established at the Mexican Red Cross command post provided the initial accurate picture of Acapulco's devastation to government authorities and relief organizations nationwide. Before this link was established, response planning was based largely on speculation and incomplete information.

Enabling coordination between emergency services. Fire departments, medical teams, search and rescue units, and law enforcement all depend on reliable communications to coordinate their efforts. Without telecommunications infrastructure, these services operate in isolation, often duplicating efforts in some areas while leaving others completely unserved. The strategic deployment of emergency communications systems at key coordination points creates the nervous system that allows the broader emergency response to function as a cohesive organism rather than disconnected parts.

Facilitating resource allocation decisions. In the aftermath of a disaster, resources are always limited relative to needs. Effective telecommunications enable authorities to gather accurate information about conditions across the affected area, prioritize interventions, and direct resources where they will have the greatest impact. Our satellite connections at emergency operations centers during Hurricane Otis enabled commanders to receive field reports from throughout the disaster zone, creating a comprehensive operational picture that informed critical resource deployment decisions.

Reconnecting communities and families. Beyond the operational aspects of disaster response, telecommunications services provide a profound psychological benefit by reconnecting affected individuals with family members and support networks. The ability to make a brief call saying "I'm alive" or send a text message to loved ones reduces anxiety and provides emotional support that sustains people through the challenging recovery period. During our Otis response, the community connectivity points we established in public squares became gathering places where residents could briefly connect with the outside world, often breaking down in tears of relief after confirming their family's safety.

Supporting recovery and reconstruction efforts. As the immediate emergency phase transitions to recovery, telecommunications infrastructure continues to play a vital role in coordinating rebuilding efforts, processing aid applications, restoring economic activities, and returning communities to a state of normalcy. The temporary communications solutions deployed during the emergency phase often become critical bridges that support recovery operations for weeks or months until permanent infrastructure can be rebuilt.

Given these critical functions, the rapid restoration of telecommunications following a disaster isn't merely a business continuity concern for service providers—it's a fundamental humanitarian need that enables all other aspects of emergency response and recovery. This understanding transforms how we approach telecommunications emergency response, elevating it from a technical challenge to an essential life-supporting service.

Personal Account: Hurricane Otis in Guerrero

I was at my office watching the weather reports with growing concern as Hurricane Otis rapidly transformed. In a matter of hours, it grew from a Category 1 to a Category 5 hurricane, defying typical development patterns and leaving little time for standard preparation protocols.

As the gravity of the situation became clear, I didn't wait for formal activation orders or complete information. My experience in both emergency medicine and telecommunications had taught me that in rapidly evolving disasters, immediate action often makes the critical difference. My first call went to the Mexican Red Cross Disaster Coordinators. As a Board Member, I had established relationships with key emergency management personnel—relationships that would prove invaluable in the critical hours ahead.

The company I was working for at that time, Hughes, immediately offered to deploy installation teams directly to Acapulco, where the hurricane would make landfall. This rapid decision-making, before the full impact was even known, exemplifies the proactive mindset required in emergency response. Rather than waiting for damage assessments and formal requests, we mobilized on the likelihood of need, knowing that communications would almost certainly be compromised.

We reached Acapulco within hours, traveling alongside early response teams and bringing critical satellite communication equipment. What we found was devastation beyond what news reports could convey—infrastructure destroyed, roads blocked, and a city almost completely cut off from the outside world.

The value of pre-established partnerships became immediately apparent. The Mexican Red Cross had deployed with mobile emergency power supplies—essential equipment in an environment where commercial power had completely failed. This partner coordination allowed us to immediately establish connectivity for the Red Cross command post, creating the first reliable communication link in the disaster zone.

Our installation teams worked around the clock, often in dangerous conditions and amid continuing bad weather, to deploy additional VSAT terminals. These efforts extended critical connectivity to the Marine units that had arrived, other emergency services coming from Mexico City, and the humanitarian aid distribution centers established by the Mexican Red Cross.

The impact of these early communication links cannot be overstated. They enabled coordination of search and rescue operations, facilitated the distribution of emergency supplies, and allowed responders to report critical needs back to command centers in Mexico City. Each VSAT installation became a lifeline not just for the emergency responders using it, but for the communities they were serving.

The Two Types of Heroes in Disaster Response

Throughout my career in emergency response, I've witnessed countless acts of heroism. But during Hurricane Otis, I observed firsthand a profound truth about disaster response that often goes unrecognized: there are two distinct types of heroes whose combined efforts make recovery possible.

The first category includes those we traditionally recognize as heroes—the search and rescue teams pulling survivors from collapsed buildings, the medical personnel working tirelessly in makeshift clinics, the Red Cross volunteers distributing essential supplies to desperate communities. Their courage is visible, their impact immediate and tangible. We see them on news reports, carrying children through floodwaters or providing comfort to those who have lost everything. Their heroism is undeniable and rightfully celebrated.

Yet there exists a second category of heroes whose contributions, while less visible, are equally critical to saving lives and restoring communities. These are the telecommunications installers and technicians who work behind the scenes to reestablish the communication infrastructure that enables all other response efforts to function effectively.

In Acapulco, I watched these technical professionals scale damaged structures in dangerous conditions, work through day and night under relentless pressure, and solve complex technical problems with limited resources. They established satellite connections in sweltering heat that regularly exceeded 100°F, often working 18-hour shifts without adequate food or water. They slept in their vehicles or on the floor of damaged facilities, foregoing basic comfort to maintain continuous operations. Many had left their own families, not knowing when they would return home.

What makes their contribution heroic isn't just the difficult conditions they endure, but the transformative impact of their work. Our team's first installation provided satellite internet connectivity to the Mexican Red Cross command post. This single connection broke the information blackout that had descended over Acapulco. Before this link was established, the rest of the country had virtually no accurate information about the true magnitude of the disaster. This wasn't merely a technical achievement—it was the critical first step in mobilizing an appropriate national response. Government agencies, relief organizations, and emergency services had been operating in an information vacuum, unable to grasp the full scale of destruction and human suffering.

Once that initial connection was established, accurate information about the disaster's true scope began to flow outward. Images and reports revealed the unprecedented devastation, catalyzing a more robust and appropriate response from federal authorities and international aid organizations. This single satellite link, established through the relentless work of technicians who had been operating continuously since the disaster struck, likely saved countless lives by ensuring that aid was properly scaled and directed to the areas of greatest need.

We select our installers and field personnel specifically for their commitment to serving their communities. During regular operations, these professionals already work under difficult circumstances, trekking to remote locations, climbing towers in harsh weather, and solving complex technical problems with limited resources as they fight to reduce the digital divide in underserved regions. For these dedicated individuals, working under disaster circumstances represents an extension of their everyday mission rather than a radical departure. Their familiarity with challenging conditions and their deep connection to the communities they serve make them uniquely suited for emergency response.

The Emergency Response Mindset in Telecommunications

This book adapts proven emergency response principles to the specialized needs of telecommunications disaster recovery. While restoring communications infrastructure differs in many ways from emergency medical response, the fundamental mindset shares important similarities:

Speed is essential. Just as the "golden hour" concept in trauma medicine recognizes that rapid intervention dramatically improves survival rates, the first 24-48 hours after a communications blackout represent a critical window during which restoring even basic connectivity can significantly enhance overall emergency response effectiveness. Every hour earlier that communications can be established means more effective coordination, better resource allocation, and ultimately, more lives saved.

Systematic protocols outperform improvisation. While creativity and adaptability remain essential in disaster response, having established protocols for common scenarios provides a crucial foundation for effective action under extreme pressure. Just as emergency medical teams follow clear procedures for assessing and treating patients, telecommunications response teams need structured approaches to assessment, prioritization, and deployment that can be executed consistently even in chaotic environments.

Team structure and clear roles matter. Emergency medical teams function with clearly defined roles and responsibilities—from triage nurses to trauma surgeons to support personnel. Similarly, effective telecommunications emergency response requires specialized teams with distinct functions, clear reporting relationships, and established coordination mechanisms. This structure enables complex operations to function efficiently despite challenging conditions and limited resources.

Resource management determines sustainability. In both emergency medicine and telecommunications response, the management of limited resources—whether medical supplies or generator fuel—often determines how many people can be helped. Systematic approaches to resource allocation, conservation, and resupply enable sustained operations beyond the initial response phase.

Documentation supports improvement. Despite the urgency of emergency response, both medical teams and telecommunications responders must maintain appropriate documentation of their actions, decisions, and outcomes. This documentation supports not only operational continuity but also post-event analysis that drives continuous improvement in response capabilities.

By adopting these principles from emergency response and adapting them to the specific requirements of telecommunications restoration, we create a framework that enables more effective, reliable, and impactful disaster response capabilities.

Who This Guide Is For

This guide is designed primarily for telecommunications organizations seeking to develop or enhance their emergency response capabilities. This includes:

Telecommunications service providers of all sizes, from national carriers to regional and local operators across all technology platforms including cellular, satellite, fixed wireless, fiber, and traditional wireline services.

Emergency management agencies responsible for coordinating disaster response, who need to understand the specific requirements and capabilities of telecommunications restoration teams.

Government telecommunications authorities charged with regulatory oversight and coordination of emergency communications during disasters.

Humanitarian organizations that increasingly depend on reliable communications for coordinating relief efforts in disaster zones.

Corporate telecommunications departments within organizations that maintain critical communications requirements during emergencies.

The guidance provided is scalable to organizations of different sizes and resources. While some recommendations may require significant investment, many of the core principles and approaches can be implemented by even small providers with limited resources. The focus throughout is on practical, field-tested approaches rather than theoretical frameworks.

How to Use This Guide

This guide is designed to be both a planning resource for developing emergency response capabilities and an operational reference during actual emergencies. Different sections will be relevant at different times in your emergency preparedness journey:

For organizations building new emergency response capabilities, the early chapters on team structure, equipment planning, and training programs provide a comprehensive framework for developing these capabilities from the ground up. Start at the beginning and work through systematically, adapting the recommendations to your specific context and resources.

For organizations enhancing existing capabilities, the assessment tools, training programs, and case studies offer opportunities to identify gaps and strengthen specific aspects of your emergency response system. Use the chapter index to focus on areas most relevant to your current needs.

For teams responding to active emergencies, the protocol sections and operational checklists provide immediately applicable guidance. These sections are designed for quick reference during crisis situations when time for reading is limited. The appendices contain practical tools and templates that can be reproduced and used in the field.

Throughout the guide, you'll find:

Case studies from actual disaster responses, particularly Hurricane Otis, that illustrate how principles and protocols function in real-world conditions.

Field-proven protocols that have been tested in actual disaster environments and refined based on experience.

Checklists and templates that can be adapted to your organization's specific needs and used for planning, training, and emergency operations.

The most effective use of this guide involves not just reading but implementing—testing the recommended approaches during training exercises, adapting them to your specific organizational context, and continuously refining them based on your experience. Remember that emergency response capabilities require ongoing maintenance and development; they cannot be established once and then left untested until disaster strikes.

In the chapters that follow, we'll explore every aspect of building, maintaining, and deploying an effective telecommunications emergency response capability. Whether you're establishing a new program or enhancing an existing one, these field-tested approaches will help your organization become a crucial part of effective disaster response—restoring the communications lifelines upon which all other emergency services depend.

MARCOS DUARTE RIVERO

CHAPTER 1: THE TELECOMMUNICATIONS EMERGENCY RESPONSE TEAM MODEL

When a multi-vehicle accident occurs on a highway, emergency medical services don't respond with a single ambulance and a generalized approach. They deploy a structured team with clearly defined roles, specialized equipment, established protocols, and a unified command system. The trauma team leader coordinates the efforts of paramedics, emergency medical technicians, and specialists, each performing specific functions within a cohesive whole. Their actions follow established protocols, but with the flexibility to adapt to the unique circumstances of each emergency.

The telecommunications industry can and should adopt a similar model for disaster response. Rather than relying on ad hoc teams hastily assembled when disaster strikes, telecom companies need dedicated emergency response teams structured around specialized functions, equipped with appropriate tools, guided by clear protocols, and led by experienced crisis managers.

The Medical-Telecom Emergency Parallel

The parallels between emergency medicine and telecom emergency response are striking and instructive. In emergency medicine, the triage nurse rapidly assesses incoming patients, categorizing them based on the severity of their condition and their chance of survival with timely intervention. Similarly, telecom emergency teams must rapidly assess damaged infrastructure, categorizing restoration priorities based on the criticality of the services supported and the technical feasibility of restoration.

Emergency medical teams operate with a clear chain of command, typically using the Incident Command System (ICS) to ensure coordination across multiple agencies and responders. This same structure applies directly to telecom emergency response, where coordination between technical teams, government agencies, and other service providers becomes essential for effective response.

Medical teams prepare extensively through training, simulations, and regular drills, ensuring that when crisis strikes, responses become almost automatic despite the unique aspects of each emergency. Telecom emergency teams must similarly engage in regular exercises, from tabletop scenarios to full-field deployments, building both technical competence and team cohesion before they're tested in an actual disaster.

Perhaps most significantly, emergency medicine understands the concept of the "Golden Hour"—the critical period immediately following traumatic injury during which prompt medical attention can significantly increase survival rates. In telecommunications restoration, we face a similar imperative. The first 24-48 hours after a communications blackout represent our own version of the Golden Hour—a critical window during which restoring even basic communications capabilities can dramatically improve overall emergency response effectiveness and save lives.

During Hurricane Otis, we witnessed this principle in action. By prioritizing the rapid deployment of satellite communications for emergency coordination centers within the first 24 hours, we enabled rescue operations that would otherwise have been dangerously uncoordinated. Every hour earlier that we could establish these communication nodes meant more effective search patterns, better resource allocation, and ultimately, more lives saved.

Core Principles of Telecom Emergency Response

Building on the medical emergency model, effective telecom emergency response rests on six fundamental principles that should guide team development, training, and operational planning.

Speed of deployment stands as the foremost principle. In the aftermath of a disaster, the value of communications restoration diminishes with every passing hour. Teams must be capable of mobilizing within minutes of activation, with pre-packed equipment, clear deployment procedures, and transportation already arranged. During Hurricane Otis, our advance team was en route within 90 minutes of the initial call, with satellite equipment and technical specialists moving toward the impact zone even before the full extent of the damage was known. This rapid deployment allowed us to establish the first emergency communications link in Acapulco within hours of gaining access to the area.

Scalability of resources forms the second critical principle. Disasters vary enormously in scale and impact, requiring response teams to scale their deployment appropriately. This means having the capability to expand from a small advance team to a full-scale operation involving dozens of technicians and tons of equipment. The organizational structure must accommodate this expansion without losing coordination or effectiveness. After our initial assessment in Guerrero, we were able to scale our response from two advance teams to seventeen specialized units across the impact zone, each with distinct missions but operating under unified command.

Self-sufficiency in the field becomes essential when responding to catastrophic events where local infrastructure has collapsed. Teams must deploy with everything needed to operate independently for at least 72 hours, including food, water, shelter, power, and security. During Hurricane Otis, our teams carried not only technical equipment but complete self-sufficiency kits that prevented them from becoming a burden on local resources already stretched to breaking point. This principle extends beyond physical needs to technical self-sufficiency—teams must have the tools, parts, and expertise to solve problems without external support.

Adaptability to changing conditions separates truly effective emergency teams from those that struggle when plans meet reality. Disaster zones are inherently unpredictable environments where conditions can change rapidly. Buildings deemed safe can become hazardous. Weather can deteriorate. Security situations can evolve. The emergency team must be capable of rapidly reassessing situations and adjusting operations accordingly. In Acapulco, when our initial plan to restore a key cellular tower proved impossible due to structural damage, the team immediately pivoted to deploying a temporary cell-on-wheels (COW) solution, losing minimal time despite the complete change in approach.

Coordination with other emergency services recognizes that telecommunications restoration never happens in isolation. It is one component of a broader emergency response involving government agencies, military units, medical teams, and other essential services. Telecom emergency teams must integrate seamlessly with these other responders, understanding their needs and priorities while effectively communicating the capabilities and limitations of emergency telecommunications solutions. During our Hurricane Otis response, embedding liaison technicians directly with emergency management agencies proved invaluable for ensuring our restoration priorities aligned with overall response objectives.

Prioritization of critical communication needs acknowledges that during the initial phase of disaster response, it's impossible to restore all services to all users. Difficult decisions must be made about which facilities, agencies, and areas receive priority attention. This prioritization should follow established frameworks that place emphasis on communications supporting life-saving operations, emergency coordination, and public safety information. In Guerrero, we implemented a tiered priority system that first established communications for emergency operations centers, then hospitals and first responder bases, followed by government coordination sites, before addressing general population needs through community connectivity points.

These six principles form the foundation of effective telecom emergency response. They inform everything from team structure and equipment selection to training programs and operational protocols. When properly implemented, they enable telecommunications teams to function as true emergency responders rather than merely technical repair units, dramatically increasing their effectiveness during the critical early phase of disaster response.

Case Study: Hurricane Otis First 72 Hours

The theoretical principles of telecom emergency response were put to their ultimate test when Hurricane Otis made landfall near Acapulco, Mexico on October 25, 2023. What followed provides a vivid case study in both the challenges of catastrophic communications failure and the effectiveness of a structured emergency response.

Otis intensified with unprecedented speed, transforming from a tropical storm to a Category 5 hurricane in less than 24 hours. This left no time for preventive measures such as deploying emergency generators or relocating critical equipment. When the hurricane struck with sustained winds of 165 mph, it devastated the telecommunications infrastructure across Guerrero state. Of 246 cell towers in Acapulco, 203 were either destroyed or severely damaged. Fiber optic backhaul was severed in multiple locations. Power infrastructure, including the main electrical substations, was extensively damaged. Even satellite communications were initially hampered by the continuing severe weather conditions.

Our emergency response began while the hurricane was still raging, with remote assessment and mobilization of resources. The initial challenge we faced was simply gathering accurate information about the extent and nature of the damage. With normal communications channels inoperative, we relied on a combination of satellite imagery, emergency radio reports, and information from government agencies to build our preliminary assessment. This highlighted the first lesson of the response: the critical importance of maintaining alternative assessment capabilities when normal channels fail.

As our advance teams approached Acapulco in the immediate aftermath, they encountered severe logistical challenges that would characterize the entire operation. Roads were blocked by debris, flooding, and landslides. Fuel was scarce. Local transportation infrastructure, including the airport, was severely damaged. These conditions required our teams to improvise access routes, bringing in initial equipment via off-road vehicles and even, in some cases, on foot. The second key lesson emerged: emergency teams must be prepared for extreme access challenges and cannot rely on normal transportation infrastructure.

Upon reaching the impact zone, our technical scouts conducted rapid assessments of key telecommunications sites. These assessments revealed the extent of the catastrophe—physical infrastructure had not merely been damaged but in many cases completely destroyed. Power systems were offline across the region with no estimate for restoration. The immediate implication was clear: temporary, self-powered communications solutions would be essential, as repair of permanent infrastructure would take weeks or months.

Based on this assessment, we made the critical decision to prioritize deployment of satellite-based communications systems to key emergency management locations. Within the first 24 hours, we established satellite internet connections at the main emergency operations center, three hospitals, and the temporary government coordination center. These connections enabled the coordination of search and rescue operations, medical evacuations, and initial aid distribution. This early deployment decision proved to be among the most consequential of the entire operation, enabling countless critical operations when every hour mattered.

As additional teams and equipment arrived during the second and third day, we expanded our response to include:

1. Deployment of multiple Cells on Wheels (COWs) in strategic locations to provide cellular coverage for emergency responders

2. Establishment of emergency community connectivity points where civilians could briefly connect to contact family members or access emergency information
3. Installation of microwave temporary backhaul to reconnect surviving cell sites to the network
4. Provision of satellite phones to key emergency officials to ensure command and control capabilities
5. Deployment of Wi-Fi hotspots at aid distribution centers to support logistics coordination

Throughout this period, our operations were guided by a clear prioritization framework that directed resources toward life-saving communications first, operational coordination second, and general public communications third. This framework, established before the disaster and understood by all team members, enabled rapid decision-making without constant escalation to leadership.

The response was not without significant challenges. Security concerns emerged in some areas as desperation grew among the population. Power for our temporary solutions required constant fuel resupply in an environment where fuel was extremely scarce. Technical failures under harsh environmental conditions required creative field repairs. Team fatigue became a factor as the operation extended beyond the initial 72 hours with limited relief options.

Perhaps the most significant challenge was managing expectations among both government officials and the general public. The comprehensive restoration of communications would take weeks, but the immediate need was perceived as absolute. This required careful communication from our liaison team about capabilities, limitations, and realistic timeframes.

By the end of the first 72 hours, we had established emergency communications capabilities that, while limited, provided the essential connectivity needed for the broader disaster response to function effectively. The operation demonstrated both the value of a structured emergency response team and the limitations that even the best-prepared teams face in catastrophic conditions.

From this intense experience, several crucial lessons emerged that have shaped our approach to telecom emergency response:

1. The value of rapid deployment of even limited communications capabilities far outweighs waiting for more comprehensive solutions.

2. Flexibility in technical approach—being willing to abandon preferred solutions in favor of what can be implemented quickly—is essential.

3. Self-sufficiency, particularly in power generation and fuel storage, determines how long emergency solutions can remain operational.

4. Coordination with government authorities, military units, and other emergency services multiplies the effectiveness of technical teams.

5. Pre-established prioritization frameworks, understood and accepted by all stakeholders before disaster strikes, prevent costly delays in decision-making during the response.

These lessons from the Hurricane Otis response inform many of the recommendations throughout this book. They represent hard-won insights that can help other telecommunications providers build more effective emergency response capabilities—capabilities that can make the critical difference when the next disaster strikes and communications fall silent.

Building the Emergency Response Mindset

Effective telecommunications emergency response requires more than just technical skills and appropriate equipment. It demands a specific mindset—a way of thinking and approaching problems that differs significantly from normal telecommunications operations. This mindset must be consciously developed and reinforced through training, simulation, and leadership.

Calmness under extreme pressure forms the cornerstone of the emergency response mindset. When critical communications infrastructure has failed and multiple stakeholders are demanding immediate restoration, the ability to remain calm and methodical is essential. This doesn't mean being unaware of the urgency—quite the opposite. It means recognizing that a clear mind makes better decisions faster, while panic leads to errors and delays. During Hurricane Otis, our most effective team leaders were those who maintained composed demeanors despite the chaos around them, providing stability that enabled their teams to function effectively.

Decisive action with incomplete information distinguishes effective emergency responders. In disaster situations, waiting for complete information before acting is rarely possible. The telecommunications emergency responder must become comfortable making decisions based on the limited information available, understanding that some assumptions will prove incorrect and plans will require adjustment. The willingness to act despite uncertainty—to deploy teams even before comprehensive damage assessments are available—often determines whether communications can be restored within the critical early window when they will have the greatest impact.

Systematic prioritization becomes essential when faced with overwhelming needs and limited resources. The emergency response mindset includes the ability to make difficult triage decisions—determining which communications needs will be addressed first, which must wait, and in some cases, which cannot be met during the initial emergency phase. This requires both a clear framework for prioritization and the psychological fortitude to implement those priorities even when faced with urgent requests that fall outside them. During our Otis response, maintaining our focus on emergency services communications despite intense pressure to restore public networks demonstrated this principle in action.

Improvisation within structure balances the need for consistent protocols with the reality that each disaster presents unique challenges. The effective emergency responder works within established frameworks but adapts them to the specific situation at hand. This requires deep familiarity with both the protocols and their underlying principles, allowing responders to determine when deviation is necessary while maintaining alignment with core objectives. Our teams in Acapulco frequently improvised technical solutions using unexpected materials and approaches, yet always remained aligned with our established prioritization framework and coordination protocols.

Endurance and self-care recognize that emergency response is a marathon, not a sprint. The telecommunications emergency responder must pace themselves and their team for sustained operations under difficult conditions. This includes attention to basics like hydration, nutrition, and rest, even when the natural impulse is to work continuously. Leaders must model appropriate self-care and enforce rest periods to maintain team effectiveness over extended operations. Following 36-hour initial deployments during Otis, we implemented mandatory 8-hour rest periods for all personnel, even when this temporarily slowed some operations, recognizing that exhausted technicians make dangerous mistakes.

Developing this mindset requires dedicated training that goes beyond technical skills to include psychological preparation, decision-making exercises, and stress inoculation. It must be reinforced through realistic simulations that reproduce the pressure, uncertainty, and resource constraints of actual disasters. Most importantly, it must be modeled by leadership at all levels of the organization, demonstrating that this approach is valued and expected when emergencies occur.

The emergency response mindset doesn't replace technical expertise—it makes that expertise effective under the most challenging conditions. By developing this mindset throughout your telecommunications emergency response team, you create the foundation for effective action when communications infrastructure fails and communities depend on your ability to reconnect them during their greatest time of need.

CHAPTER 2: BUILDING YOUR EMERGENCY RESPONSE TEAM

A telecommunications emergency response team is more than a collection of technical experts. It is a carefully structured organization where each role has been defined with specific responsibilities, reporting relationships, and skill requirements. The structure must balance specialization with flexibility, technical expertise with operational capabilities, and field deployment with strategic coordination.

Drawing from both emergency medical team models and our experience in telecommunications disaster response, this chapter outlines how to build a team that can function effectively under the most challenging conditions. The structure presented here can be scaled according to the size of your organization, though the fundamental components remain consistent whether you're a regional provider or a national telecommunications corporation.

Team Structure and Roles

An effective telecom emergency response team consists of four major functional groups, each with distinct responsibilities yet designed to work in close coordination. This structure mirrors the specialized-yet-integrated approach used in emergency medicine, where trauma teams include specialists focused on different aspects of patient care but functioning as a unified team.

Field Assessment Teams

Field Assessment Teams serve as the eyes and ears of your emergency response operation. Similar to how paramedics perform initial patient assessment before treatment begins, these teams rapidly evaluate the situation on the ground, providing the critical intelligence that will guide all subsequent deployment decisions.

Technical scouts form the core of these teams, typically comprised of experienced field engineers with broad knowledge across multiple telecommunications technologies. These individuals must be able to quickly assess damage to various infrastructure components—from cell towers and microwave links to fiber optic cables and power systems. During the Hurricane Otis response, our technical scouts provided the first accurate assessment of tower damage in Acapulco, contradicting early reports and enabling more effective resource allocation.

Infrastructure evaluators focus specifically on the structural integrity of telecommunications facilities. These specialists, often with civil or structural engineering backgrounds, determine whether existing structures can be repaired or must be bypassed with temporary solutions. Their assessments prevent dangerous deployment of personnel to compromised structures and identify where temporary infrastructure can be safely installed. In Guerrero, rapid structural assessment of key telecommunications buildings prevented teams from entering dangerously unstable facilities, potentially avoiding serious injuries.

Environmental risk assessors evaluate ongoing or potential hazards in the disaster zone that could affect both the safety of response teams and the viability of emergency solutions. These specialists identify hazards from unstable ground, flood risks, hazardous materials, and continuing weather threats. Their work ensures that emergency solutions are not deployed in locations where they might soon be compromised by evolving environmental conditions. During Hurricane Otis recovery, environmental assessments identified several areas with continuing landslide risk, redirecting deployments to more stable locations.

Security analysts evaluate the safety situation in affected areas, identifying regions where emergency teams can operate safely and those requiring additional security measures or coordination with law enforcement. In disaster zones, normal social order can be disrupted, creating potential security concerns that must be addressed before technical teams deploy. Following Hurricane Otis, our security analysts worked closely with military and police units to ensure our teams could safely access all required areas, particularly during night operations.

Field Assessment Teams are typically the first to deploy following a disaster, moving quickly into affected areas to gather the critical information that will guide the entire response effort. They must be highly mobile, self-sufficient, and capable of operating in austere conditions with minimal support. Their equipment is focused on assessment rather than restoration, including specialized diagnostic tools, structural evaluation equipment, satellite communication devices, and comprehensive documentation capabilities.

Rapid Deployment Units

While Field Assessment Teams gather information, Rapid Deployment Units deliver immediate communications solutions to the most critical locations. These specialized teams are the equivalent of emergency medical intervention—they provide the immediate, life-saving capabilities needed to stabilize the situation until more comprehensive restoration can begin.

Satellite communications specialists form the tip of the spear in most major telecommunications disasters. These technicians deploy and operate various satellite-based communication systems, from portable VSAT terminals to emergency satellite phones and backpack satellite units. Their expertise includes not just technical setup but optimization of limited satellite bandwidth for maximum effectiveness. In Acapulco, our satellite specialists established the first emergency communications link at the main hospital within hours of gaining access, enabling coordination of medical evacuations that saved numerous lives.

Temporary tower deployment technicians specialize in the rapid installation of portable cellular infrastructure, including Cells on Wheels (COWs), Cells on Light Trucks (COLTs), and portable base stations. These highly specialized teams can establish temporary cellular coverage in areas where permanent infrastructure has been damaged or destroyed. Their work requires not just technical expertise but physical capabilities to set up this equipment in challenging environments. During the Hurricane Otis response, these teams deployed five COWs across the Acapulco region within 48 hours, restoring cellular coverage for emergency responders across approximately 60% of the critical operational areas.

Power restoration teams focus exclusively on establishing and maintaining electrical power for emergency communications equipment. In major disasters, commercial power is often unavailable for extended periods, making self-generated power essential for telecommunications restoration. These specialists deploy and operate generators, solar systems, and battery arrays, while managing the complex logistics of fuel supply in disaster zones. Their work is often the limiting factor in how long emergency communications can be maintained. Following Hurricane Otis, our power teams maintained continuous operation of emergency equipment despite widespread fuel shortages by implementing careful conservation measures and establishing dedicated supply chains.

Mobile communications center operators manage deployable command and technical operations centers that provide both coordination capabilities and technical support for field teams.

These sophisticated units serve as hubs for emergency operations, often including satellite connectivity, technical repair facilities, planning spaces, and rest areas for technical staff. Operating these complex systems requires specialists familiar with all aspects of their functionality. Our mobile command center in Acapulco became the hub of operations for the entire region, providing continuous coordination despite the challenging conditions.

Rapid Deployment Units constitute the main operational force of your emergency response team. They deploy immediately following initial assessment, bringing with them the equipment and expertise needed to establish emergency communications in the most critical locations. These units must be completely self-sufficient, capable of operating independently in disaster zones for 72-96 hours without external support. Their training emphasizes both technical proficiency and operational resilience under extreme conditions.

Command and Control Center

While field teams manage the tactical response, the Command-and-Control Center provides strategic direction, resource coordination, and external communication. This function mirrors the incident command structure used in emergency medicine, ensuring coordinated effort across multiple teams operating in challenging conditions.

The Emergency Operations Director serves as the overall leader of the response effort, making critical strategic decisions about resource allocation, deployment priorities, and risk management. This individual, typically a senior leader with both technical understanding and emergency management experience, maintains the broader operational perspective while field teams focus on specific missions. The director is ultimately accountable for the effectiveness of the entire response operation. During Hurricane

Otis, our operations director maintained continuous oversight of seventeen separate deployment teams operating across a wide geographic area, ensuring their efforts remained synchronized despite complex and changing conditions.

The Resource Allocation Manager oversees the distribution of personnel, equipment, and supplies across the disaster area. This crucial role requires detailed tracking of all available resources, understanding of their capabilities and limitations, and the ability to match resources to needs as identified by assessment teams. The resource manager also coordinates logistics for getting additional resources to the disaster zone as the operation continues. In Guerrero, effective resource allocation enabled us to support simultaneous operations across multiple communities with limited equipment, maximizing the impact of our response.

The Interagency Liaison coordinates with government agencies, military units, other telecommunications providers, and additional emergency services. This role ensures that telecommunications restoration efforts align with broader emergency response priorities and that resources can be shared effectively when appropriate. The liaison also manages permissions, access authorizations, and regulatory compliance during emergency operations. During Hurricane Otis, while our Hughes team focused specifically on establishing the initial four critical satellite connections for the Mexican Red Cross, we observed how major carriers deployed liaison teams that embedded directly with federal emergency management authorities. These liaison efforts by the larger providers ensured their restoration priorities matched the government's broader response strategy. Our focused experience with the Red Cross installations gave us valuable insight into how these coordination mechanisms functioned across the broader emergency response ecosystem.

The communications Coordinator manages information flow both within the emergency response team and to external stakeholders. Internally, this role ensures field teams receive critical updates, changing priorities, and safety information. Externally, the coordinator provides accurate information to government officials, other telecommunications providers, and when appropriate, the media and public. This role prevents the spread of misinformation that can hamper effective response. Following Hurricane Otis, our communications team provided twice-daily briefings to government authorities and partner organizations, ensuring aligned understanding of the telecommunications situation across all responding agencies.

The Logistics Supervisor manages the complex supply chains needed to sustain emergency operations in disaster zones. This includes transportation, accommodation when available, food and water, fuel supplies, equipment replacement, and medical support for team members. In major disasters, logistics often become the determining factor in how long effective operations can be maintained. During our Acapulco operations, the logistics team established a continuous supply line from Mexico City to support field teams, including dedicated fuel transport for generators when local supplies were exhausted.

The Command-and-Control Center may operate from a fixed location away from the disaster zone or from a mobile command post deployed to the affected area, depending on the nature and scale of the emergency. In either configuration, it must maintain reliable communications with all field teams while having the systems and staff needed to coordinate complex operations. The center operates on a 24/7 basis during active emergencies, typically with rotating staff to ensure continuous, effective oversight of all activities.

Support Functions

Beyond the operational teams and command structure, several specialized support functions are essential for sustained emergency response. These roles may not be deployed directly to the disaster zone but are nonetheless critical to the success of the overall operation.

Transportation and access specialists focus on solving the often-complex challenges of moving personnel and equipment into and around disaster zones. These experts work with aviation units for helicopter transport, coordinate with maritime resources for coastal access, arrange for specialized off-road vehicles, and navigate the bureaucratic requirements for emergency vehicle authorizations. During the Hurricane Otis response, our transportation specialists arranged for helicopter delivery of critical equipment to otherwise inaccessible mountain sites where key telecommunications infrastructure was located.

Supply chain managers ensure the continuous flow of equipment, parts, fuel, and supplies needed to sustain operations. They coordinate with vendors, manage emergency procurement processes, track inventory, and prioritize deliveries based on operational needs. In extended operations, the effectiveness of the supply chain becomes increasingly important as initial resources are consumed. Following Hurricane Otis, our supply chain team activated emergency procurement protocols that delivered over 200,000 liters of generator fuel to the disaster zone despite widespread shortages.

Staff support and rotation planning specialists manage the human dimension of emergency response, including shift scheduling, rest requirements, team rotations, and support for the physical and psychological needs of responders. These functions become especially critical in extended operations where fatigue and stress can significantly impact performance. During our operations in Guerrero, this team implemented a rotation system that brought fresh personnel into the disaster zone after the first week, maintaining operational effectiveness as the response extended beyond initial deployment.

Administrative documentation specialists manage the critical but often overlooked task of documenting all aspects of the emergency response. This includes tracking expenses for potential reimbursement, maintaining records of decisions and actions for after-action review, and collecting the data needed for insurance claims and regulatory reporting. In the Hurricane Otis response, comprehensive documentation enabled full cost recovery through insurance and government emergency programs, while also providing valuable data for improving future responses.

Support functions typically operate from outside the immediate disaster zone but maintain close coordination with both field teams and the command center. Their work is less visible than front-line operations but often determines whether a response can be sustained beyond the initial deployment phase. Effective emergency response requires recognizing the critical nature of these support roles and providing them with the resources and authority needed to fulfill their responsibilities.

Selection Criteria for Team Members

Not everyone with technical skills is suited for emergency response work. The challenging conditions, high pressure, and unique demands of disaster operations require specific personal qualities beyond technical expertise. Selecting the right personnel for your emergency response team is perhaps the most critical factor in its ultimate effectiveness.

Technical qualifications naturally form the foundation for team selection. Team members must possess the specialized skills required for their specific roles, whether satellite communications, cellular network deployment, power systems, or structural assessment. These qualifications should be verified through certification, demonstrated experience, and practical skills assessment rather than simply reviewing résumés. During our team development, we implemented practical field exercises that tested technical skills under realistic conditions, revealing capabilities that weren't apparent from paper qualifications alone.

Emergency experience, either in telecommunications or other emergency response roles, provides valuable perspective that cannot be taught quickly. Team members with prior disaster response experience understand the realities of working in compromised environments, have realistic expectations about conditions, and have developed the adaptive mindset needed for effective response. When building our team, we specifically recruited individuals with backgrounds in military service, emergency management, and international disaster response to complement those with purely telecommunications experience.

Physical capabilities must match the demands of emergency deployment. Team members should be able to work long hours in challenging conditions, carry required equipment, function effectively in adverse weather, and maintain performance despite physical discomfort. This doesn't mean every team member needs to be in peak athletic condition, but they must have the stamina and capability required for their specific role. For our field deployment teams, we established minimum physical requirements including the ability to carry 25kg of equipment for 2km and work effectively for 12-hour operational periods.

Psychological resilience is perhaps the most important yet difficult quality to assess. Emergency responders must make critical decisions under extreme pressure, adapt to rapidly changing conditions, function effectively despite uncertainty, and maintain emotional stability when confronted with destruction and human suffering. Identifying this resilience typically requires a combination of psychological screening, scenario-based assessment, and observation during training exercises. We incorporated psychological resilience assessment into our team selection process after observing that technical experts without this quality often became ineffective in actual disaster conditions.

Decision-making under pressure distinguishes exceptional emergency responders from those who merely possess technical skills. Team members must be able to analyze complex situations with incomplete information, identify the most critical actions, and implement decisions confidently despite uncertainty. This capability can be assessed through simulation exercises where candidates face escalating challenges with limited time for response. Our selection process includes scenario-based exercises specifically designed to evaluate how candidates perform when faced with multiple, simultaneous problems under time pressure.

Team compatibility cannot be overlooked in selecting emergency response personnel. During disaster operations, teams live and work in close proximity under stressful conditions for extended periods. Individuals who cannot function effectively in team environments, regardless of their technical capabilities, can compromise the entire operation. Assessing team compatibility includes evaluating communication styles, willingness to support others, response to authority, and ability to resolve conflicts constructively. We evaluate this quality through group exercises during selection and training, identifying both positive team influences and those whose personal styles might create operational friction.

Family circumstances and availability must be considered realistically when selecting team members. Emergency responders may be deployed with minimal notice for indeterminate periods, often without reliable communication back home. Team members must have family situations that can accommodate these demands without creating unmanageable personal stress. During our selection process, we include frank discussion of these requirements with candidates and, when appropriate, with their families, ensuring everyone understands the commitments involved.

Cross-training potential represents the capacity for team members to perform multiple functions as needed during emergency operations. While specialists are essential for complex technical functions, team members who can operate effectively across several roles provide crucial operational flexibility when challenges arise. We specifically select for and develop this capability, creating a team where many members can perform at least basic functions across multiple specialties, allowing for continuous operations even when specific experts are unavailable.

The selection process for emergency response team members should be rigorous, transparent, and specifically designed to identify the unique combination of technical skills and personal qualities needed for this challenging work. It typically includes:

1. Initial screening based on technical qualifications and experience

2. Physical capability assessment appropriate to role requirements

3. Psychological evaluation focused on stress resilience

4. Scenario-based exercises simulating emergency conditions
5. Team compatibility assessment through group activities
6. Detailed discussion of availability and deployment expectations
7. Evaluation of cross-training potential across multiple functions

This comprehensive selection process requires significant investment but pays tremendous dividends when your team faces actual emergency conditions. The quality of your personnel will ultimately determine your capability to restore critical communications when communities need them most.

Personal Preparedness Requirements

Even the most qualified emergency responders cannot function effectively without proper personal preparation. The unique demands of disaster deployment require specific preparation measures that should be standardized across your emergency response team.

Go-bags and personal equipment must be maintained in a constant state of readiness by all team members designated for potential field deployment. These pre-packed kits contain everything needed for immediate deployment to disaster zones, including appropriate clothing for various conditions, personal protective equipment, hygiene supplies, medications, identification documents, and basic survival tools. Team members should be able to deploy within 30 minutes of notification using only these pre-positioned resources. Our standard requirement includes three distinct go-bags: a 24-hour pack carried at all times, a 72-hour pack maintained at home or office, and a 7-day deployment kit stored at the emergency operations center.

Family emergency plans are essential for team members who may deploy with minimal notice. These plans address how families will function during the member's absence, including childcare arrangements, bill payment, home maintenance, and communication protocols. Without confidence that their families are prepared for their absence, team members cannot focus effectively on emergency response duties. We require all team members to develop and regularly update formal family emergency plans, which are reviewed as part of our readiness certification process.

Health and medical requirements include maintaining appropriate vaccinations, regular fitness assessments, and documentation of medical conditions and medications. Team members should have recent medical clearance for emergency deployment and carry any required personal medications in sufficient quantity for extended operations. Medical documentation, including blood type, allergies, and existing conditions, should be maintained in both personal records and central team databases. Following an incident where a team member required emergency medical care during deployment, we implemented a standardized medical readiness protocol that includes quarterly health verification.

Personal communications backup systems ensure team members can maintain essential contact with both emergency operations and their families despite compromised infrastructure. This typically includes satellite phones, emergency radios, or other alternative communications devices that team members are trained to operate. These systems should be regularly tested to ensure familiarity and functionality. Our standard deployment kit includes a personal satellite communicator for each team member, with predetermined check-in protocols for both operational and family communication.

Psychological preparation involves developing the mental resilience and coping strategies needed for extended operation under stressful conditions. Team members should receive training in stress management techniques, psychological first aid, and recognition of signs of burnout or traumatic stress. Regular psychological readiness assessments should be included in team preparation activities. After observing the psychological impacts of extended disaster operations, we implemented a mandatory psychological preparation program that includes both training and regular assessment components.

Financial arrangements during deployment must be clearly established before emergencies occur. This includes understanding of compensation policies for emergency work, expense reimbursement procedures, insurance coverage during deployments, and mechanisms for handling personal financial obligations while deployed. Uncertainty about these matters can create significant distraction during operations. Our team members operate under clear emergency deployment policies that address all financial aspects of disaster response, from overtime compensation to expedition allowances for field conditions.

Personal preparedness should be regularly verified through formal readiness checks conducted on both scheduled and surprise bases. These assessments ensure that team members maintain the required state of readiness between emergencies and identify any gaps requiring correction. The results of these readiness checks should be documented and addressed through individual development plans when deficiencies are identified.

The comprehensive personal preparedness standards described here may seem excessive when compared to normal business operations. However, telecommunications emergency response is not normal business. It is specialized work conducted under extraordinary conditions, often with lives depending on the outcome. The level of personal preparedness directly impacts how quickly and effectively your team can respond when disaster strikes.

When Hurricane Otis hit with unexpected intensity, our team's personal preparedness made the difference between immediate deployment and lengthy delays. Team members activated with their pre-positioned equipment, their families functioning under established emergency plans, and their personal affairs in order. This allowed them to focus completely on the mission without the distractions that often compromise emergency effectiveness.

Field-Proven Protocols: Personal Safety Hierarchy

Our field teams follow a strict hierarchy of priorities during disaster response:

1. "First me" (the technician)
2. "Then me again"
3. "Then me once more"
4. "Then my teammate"
5. "Then my vehicle"
6. "Then my tools and equipment" (radios, modems, antennas, etc.)
7. "Finally, the installation site"

This hierarchy isn't selfish—it's practical. A technician who becomes injured, ill, or stranded cannot help restoring critical communications. During Hurricane Otis, this prioritization framework guided all personnel decisions, ensuring our teams remained effective throughout the extended operation.

Field-Proven Protocols: Personal Preparedness Kit

For deployment to disaster zones, each team member carries:

Personal Essentials:

- Regular medications (5-day supply carried on their person, not in vehicles or bags that could be lost or stolen)
- Three changes of clothing
- Chargers and fully charged power banks
- Ten liters of potable water
- Four liters of electrolyte drinks
- Non-perishable food, preferably energy bars or granola (minimum five bars per day)
- Sunscreen and insect repellent

Emergency Medical Kit (in waterproof container):

- Valid identification and emergency contact information
- Five anti-diarrheal tablets
- Four ibuprofen tablets
- Ten sterile gauze pads
- Two elastic bandages
- One bottle of antiseptic (preferably benzalkonium chloride for wound cleaning)
- Water purification tablets

- Lighter
- Flashlight with fresh batteries
- Basic backup mobile phone (inexpensive but operational)

This kit should always be accessible in the vehicle's glove compartment or in a backpack that remains with the technician at all times.

CHAPTER 3: EQUIPMENT AND RESOURCE PLANNING

When disaster strikes and telecommunications infrastructure fails, the effectiveness of your emergency response depends not only on the quality of your team but also on the tools they bring to the disaster zone. This chapter explores the critical equipment and resource planning that enables rapid, effective restoration of communications in crisis conditions.

The Mobile Communication Arsenal

The modern telecommunications emergency response team relies on a diverse array of specialized equipment that can be rapidly deployed to disaster zones. This mobile arsenal must be carefully selected, meticulously maintained, and strategically positioned for immediate deployment when needed. Unlike standard telecommunications equipment, emergency response gear must function under extreme conditions, operate independently of local infrastructure, and be transportable to challenging locations.

Satellite Communication Systems

Satellite communications form the backbone of most disaster telecommunications responses, providing connectivity when terrestrial infrastructure has failed completely. These systems offer the unique advantage of independence from local infrastructure, making them invaluable during catastrophic events.

VSAT (Very Small Aperture Terminal) deployable units represent the workhorse of emergency satellite communications. These systems, consisting of a portable satellite dish (typically 0.75-1.8 meters in diameter), a mount, and associated electronics, can be transported to disaster zones and established within 1-2 hours by experienced technicians. Modern rapid-deployment VSATs feature auto-pointing capabilities that significantly reduce setup time and technical complexity, though manual pointing remains an essential backup skill for all deployment teams.

The most effective VSAT units for emergency response share several critical characteristics. They must be packaged in rugged, transportable cases that protect sensitive components during rough transport. They should feature modular designs allowing for quick assembly without specialized tools. The most advanced units include integrated backup power systems that provide 4-6 hours of autonomous operation. Crucially, they must be field-serviceable, with common repair components included in deployment kits.

Portable satellite phones provide essential voice communications when cellular networks are completely disabled. Modern satellite phones have evolved significantly, with reduced size, better battery life, and improved voice quality. Some advanced models now include limited data capabilities suitable for text-based communications and basic email. While bandwidth constraints prevent these devices from supporting video or significant data transfer, they provide crucial command and control communications for field teams operating in areas without other connectivity options.

Backpack satellite terminals represent the newest category of emergency satellite equipment, offering impressive capabilities in extremely portable packages. These systems, weighing typically 7-15kg including batteries, can be carried by a single technician into areas inaccessible by vehicle. They provide bandwidths of 2-5 Mbps, sufficient for critical communications including video conferencing with emergency management authorities. Their compact size and rapid deployment capability make them ideal for establishing initial communications in areas where larger VSAT systems cannot be immediately transported.

When selecting satellite communications equipment for emergency response, several factors beyond pure technical specifications must be considered. Geographic coverage is paramount—different satellite constellations cover different regions, and your equipment must be compatible with satellites serving your operational areas. Licensing and regulatory compliance vary by country, and pre-established approvals are essential to avoid delays during deployment. Service contracts must include emergency priority provisions that ensure bandwidth availability during disasters when systems may become congested.

My Experience: Hughes VSAT Deployment During Hurricane Otis

During Hurricane Otis, my work with Hughes focused specifically on rapidly deploying VSAT satellite communications equipment to establish critical connectivity for the Mexican Red Cross. As I watched the hurricane's rapid intensification from Category 1 to Category 5 in my office, the decision to act quickly rather than wait for formal assessments proved crucial.

Our team prepared four complete VSAT systems, prioritizing equipment that could be quickly installed in challenging conditions. We selected auto-pointing systems where available, but also brought manually adjusted equipment with the necessary tools for field setup. Each system was pre-configured and tested before deployment to minimize setup time in the disaster zone.

Upon reaching Acapulco, we faced the sobering reality of complete communications blackout. Working alongside the Mexican Red Cross's emergency power teams, we established the first reliable satellite connection at their command post within hours of arrival.

This single link became the critical communications channel that broke the information vacuum, allowing the first accurate reports of Acapulco's devastation to reach government authorities and relief organizations.

The conditions were challenging—extreme heat, no electricity, limited working space, and the psychological pressure of knowing how much depended on our success. Yet the pre-deployment preparation and the reliability of our equipment enabled us to establish this vital connection despite these obstacles.

This experience highlighted three critical lessons about satellite equipment for emergency response:

1. Pre-configuration and testing before deployment saves crucial time in the field
2. The reliability of equipment under adverse conditions matters more than advanced features
3. The integration with emergency power sources (in our case, provided by the Red Cross) is as important as the communications equipment itself

Field Observations: Broader Communications Response

While our Hughes team focused on these critical initial installations for the Red Cross, I observed how other telecommunications providers deployed a much wider array of equipment throughout the affected area as the response expanded.

Major carriers brought in Cells on Wheels (COWs) to strategic locations around Acapulco, providing cellular coverage in areas where permanent towers had been destroyed. These trailer-mounted systems included complete cellular base stations, extendable masts, power generation, and satellite backhaul. Their deployment required careful site selection balancing coverage objectives, physical access, and security considerations.

I observed how some providers deployed microwave link equipment to reconnect surviving cell sites to their networks after fiber backhaul was severed. These systems provided high-bandwidth connections over distances of several kilometers, enabling the restoration of service to substantial areas without requiring full tower replacement.

Various organizations also utilized portable cellular repeaters in key locations like emergency medical treatment points and temporary shelters. These suitcase-sized devices amplified existing cellular signals from distant towers, extending coverage into areas with weak or no direct service.

These field observations of the broader telecommunications response demonstrated the importance of having a diverse emergency equipment portfolio ready for deployment. Different scenarios required different technical solutions, and the most effective responders were those who had prepared multiple options before the disaster struck.

Temporary Cellular Infrastructure

While satellite communications provide essential connectivity for emergency coordination, the broader public relies primarily on cellular service for communication. Restoring some level of cellular coverage quickly can significantly reduce public anxiety, enable civilian emergency calls, and support community coordination during recovery efforts. Modern temporary cellular infrastructure allows for rapid restoration of service in areas where permanent infrastructure has been damaged or destroyed.

Cells on Wheels (COWs) represent the most comprehensive temporary cellular solution. These trailer-mounted systems include a complete cellular base station, antennas typically mounted on an extendable mast (15-30 meters), power generation equipment, and backhaul connectivity options (satellite, microwave, or fiber where available). A fully equipped COW can provide cellular coverage to an area of 5-20 square kilometers depending on terrain and antenna height.

The effective deployment of COWs requires careful planning and specialized capabilities. Site selection is critical, balancing coverage objectives, physical access, security considerations, and backhaul availability. Deployment teams must be skilled in both technical configuration and the mechanical setup of these substantial systems. Rigorous preventive maintenance is essential, as these complex systems incorporate multiple potential points of failure that must be identified and addressed before deployment.

Cells on Light Trucks (COLTs) offer a more mobile but less capable alternative to COWs. Mounted on pickup trucks or small vans rather than trailers, COLTs can access more challenging locations but typically feature shorter masts (10-15 meters), less powerful transmission equipment, and more limited power generation capabilities. The reduced coverage area (typically 3-10 square kilometers) is offset by greater mobility and faster deployment times.

Portable cellular repeaters provide the most lightweight cellular restoration option. These suitcase-sized devices amplify existing cellular signals from distant towers, extending coverage into areas with weak or no direct service. While limited in capability—they require some existing signal to amplify and provide coverage to only small areas (100-500 meters radius)—they can be deployed in minutes by a single technician.

Femtocells and portable base stations represent the newest generation of emergency cellular infrastructure. These small, low-power cellular access points connect to any available internet connection (typically satellite in disaster scenarios) and provide localized cellular service within a 50–200-meter radius. While limited in coverage area, they enable standard cellular phones to connect in their vicinity, requiring no special equipment for end users.

The cellular technologies selected for emergency response must be compatible with the networks and devices prevalent in your operational area. In many regions, deploying both GSM/LTE and CDMA technologies may be necessary to serve different carrier customers. Regulatory considerations are particularly important for cellular deployments, with many countries requiring specific emergency authorizations before temporary cellular infrastructure can be activated.

Field Observations: Alternative Power Systems in Disaster Zones

During Hurricane Otis, I observed how power management became one of the most critical and challenging aspects of the telecommunications response effort. Commercial power infrastructure was among the first to fail and would clearly be among the last to be restored, making independent power generation essential for all communications solutions.

Major carriers deployed multiple generators ranging from 5-20 kW to support their emergency communications installations. These generators required sophisticated logistics for fuel supply in an environment where commercial fuel was unavailable, and roads were often impassable. I watched as teams established dedicated fuel supply chains, sometimes requiring military escort in areas with security concerns.

Some advanced deployments incorporated hybrid power systems combining diesel generators with solar arrays and battery banks. These configurations significantly extended operational capability by reducing generator runtime and fuel consumption—a critical advantage when fuel resupply was uncertain. The most sophisticated systems included intelligent power management that automatically switched between power sources based on availability and load requirements.

Perhaps the most impressive aspect of the power management I observed was the implementation of power prioritization systems. These setups automatically shed non-essential loads during power limitations, ensuring that core communications functions remained operational even when fuel supplies ran low. This capability allowed critical communications links to continue functioning during periods when fuel deliveries were delayed due to road conditions or security issues.

Through these observations, it became clear that in extended emergency operations, power management expertise is as valuable as communications technical skills. The most effective response teams included specialists dedicated exclusively to establishing and maintaining electrical power for communications equipment.

Emergency Network Infrastructure

Beyond the visible communications equipment and power systems, emergency telecommunications response requires a range of specialized network infrastructure to establish functional communications services. These components create the technical foundation that enables all other emergency communications capabilities.

Portable routers and switches form the core of emergency network infrastructure, creating local networks at deployment sites and managing connections between different communications systems. Unlike standard enterprise networking equipment, emergency-grade routers must feature extreme environmental tolerance, backup power options, remote management capabilities even over constrained connections, and simplified configuration for field deployment.

Mesh network kits enable the extension of network connectivity between nearby sites without requiring cable installations. These systems establish wireless connections between network nodes, automatically routing traffic through the most efficient available paths and adapting to changing conditions or equipment failures.

Microwave link equipment provides high-bandwidth connections over distances of 1-50 kilometers with appropriate tower height and line-of-sight conditions. These systems can deliver hundreds of megabits to multiple gigabits of throughput, enabling the interconnection of emergency sites or the reestablishment of backhaul connectivity for surviving cellular infrastructure.

Fiber splicing and repair kits enable the restoration of damaged fiber optic infrastructure when complete replacement isn't necessary. Modern emergency splice kits include portable fusion splicers, test equipment, and weatherproofing materials that allow field technicians to repair critical fiber connections even in challenging environmental conditions.

Network security and access control systems are increasingly important components of emergency telecommunications deployments. These systems ensure that limited emergency bandwidth is protected from unauthorized use, sensitive emergency communications remain secure, and network resources are allocated according to established priorities.

The key to effective emergency network infrastructure lies in pre-configuration and simplification. All components should be prepared with base configurations before deployment, requiring minimal on-site setup. Documentation must be clear and concise, enabling field technicians to implement complex network architectures under pressure. Equipment should be standardized where possible, reducing the variety of spare parts required and enabling technicians to become proficient with all deployed systems.

My Experience: Network Simplification for Rapid Deployment

During our Hughes deployments for the Mexican Red Cross in Acapulco, we experienced firsthand the value of simplified network infrastructure in emergency conditions. Our approach focused on pre-configured network settings that required minimal field adjustment, allowing us to establish functional networks within minutes of the physical VSAT installation.

Each of our systems included pre-configured routers with standardized network settings and predefined security policies. We created laminated quick-reference cards with essential network information, eliminating the need to search through documentation under difficult field conditions. This preparation allowed us to concentrate on the physical installation challenges without being slowed by complex network configuration issues.

One valuable lesson came when we needed to reconfigure a system to accommodate unexpected requirements at the Red Cross command post. The simplicity of our network design allowed us to quickly adapt the configuration without specialized tools or extensive technical procedures. This experience reinforced that in emergency deployments, simplicity and adaptability often outweigh sophisticated features that might be valuable in normal operations.

Logistics and Supply Chain Management

Even the most advanced telecommunications equipment is useless without a logistics system that can deliver it to the disaster zone, sustain its operation, and support the technical teams deploying it. Emergency logistics requires specialized capabilities that differ substantially from normal business operations.

Pre-positioned Resource Caches

The most effective emergency telecommunications responses rely on strategically positioned equipment caches that can be quickly accessed when disasters occur. These pre-positioned resources dramatically reduce response times compared to mobilizing equipment from central warehouses or vendor locations.

Organizations maintaining robust emergency response capabilities typically establish regional emergency caches distributed geographically to ensure rapid access regardless of disaster location. Each cache contains a standardized set of equipment including VSAT terminals, portable networking infrastructure, power generation systems, and the tools required for their deployment. These caches are positioned based on risk analysis of potential disaster locations, transportation accessibility, and security considerations.

The management of these caches requires disciplined inventory control, regular maintenance, and equipment rotation to ensure all components remain in deployment-ready condition. Batteries must be maintained on charge controllers, generators require regular testing and preventive maintenance, and communications equipment must receive firmware updates and configuration changes as systems evolve. Effective organizations implement quarterly verification protocols for all cache locations, with full operational testing of all equipment.

My Experience: Pre-positioned Equipment Advantage

The Hughes response to Hurricane Otis benefited significantly from pre-positioned equipment. We had maintained satellite equipment in a state of operational readiness, which allowed us to begin movement toward Acapulco within hours of making the decision to deploy. This early mobilization proved crucial to establishing critical communications during the first 24 hours after the hurricane struck.

Our pre-positioned resources included not just the VSAT equipment itself, but all the supporting components needed for rapid deployment—mounting hardware, cables, tools, and even basic spare parts for field repairs. This comprehensive approach eliminated delays that would have occurred if we had needed to gather these resources from various locations after the emergency began.

The experience reinforced my belief that emergency preparedness must include not just having equipment available, but having it packaged, tested, and positioned for immediate deployment. The time saved through this preparation directly translated to earlier establishment of critical communications links when they were most urgently needed.

Field-Proven Protocols: Vehicle Preparation

Each response vehicle should carry:

Essential Supplies:

- Sealed fuel containers with sufficient fuel to return to the base station (remember: fuel is typically unavailable in affected areas)
- Additional containers of potable water
- Spare tire, jack, and tools
- Power inverter
- Rope
- Four wooden blocks (for stabilizing vehicles in soft ground)
- Reflective warning triangles
- Fire extinguisher
- Emergency light with fresh batteries

Vehicle Identification:

- Beyond company logos, vehicles should display one of the following terms in reflective vinyl:
    - Internet
    - Telecommunications
    - Telephony
    - Television

    - Radio

- Consider adding universally recognized telecommunications symbols

Equipment Organization:

- Place food and water at the bottom of cargo area
- Position additional fuel away from food and water, keeping it concealed
- Place valuable tools (rotary hammers, lights, measuring devices) in the middle layer
- Cover everything with cables, antennas, and less valuable items
- If possible, use a company-branded tarp to cover equipment (this also provides rain protection)
- Keep the spare tire, jack, and essential vehicle tools readily accessible to avoid unloading everything in case of a breakdown

Technical Supplies:

- Bring sufficient materials for installations in areas without electrical power
- Prepare enough equipment to establish at least six connectivity sites
- Include portable power generators and battery-powered tools (drills, screwdrivers, etc.)
- Verify all equipment functions properly and is fully charged before departure

Field Observations: Transport and Access Solutions

During the broader Hurricane Otis response, I observed how transportation and access challenges often became the limiting factor in how quickly telecommunications equipment could reach affected areas. Standard transportation infrastructure was severely compromised, requiring specialized capabilities to overcome these obstacles.

Major telecommunications providers deployed a variety of specialized vehicles, including high-clearance 4x4 trucks modified with additional equipment such as winches, reinforced suspensions, and protected cargo areas. These vehicles navigated extensively damaged roadways to transport equipment and technical teams throughout the affected region.

In some cases, helicopter transport was used to reach isolated locations where road access was completely cut off. This aerial capability dramatically reduced deployment time for critical initial equipment, though the limited cargo capacity restricted what could be delivered through this method.

The most effective transportation strategies I observed involved tiered capabilities matched to specific phases of the response. Initial assessment and critical first communications deployed via helicopter or specialized off-road vehicles, while heavier equipment followed on more conventional transportation as routes were gradually cleared.

This experience highlighted the importance of developing flexible transportation capabilities and establishing relationships with specialized transport providers before disasters occur. The organizations that had pre-arranged access to various transportation modes were able to deploy communications solutions far more rapidly than those limited to conventional vehicles.

Maintenance During Prolonged Deployment

As emergency operations extend beyond the initial response phase, equipment maintenance becomes increasingly important to sustained communications capability. Extended operation in harsh environments accelerates wear on all systems, requiring proactive maintenance protocols.

Field service kits should accompany all major equipment deployments, containing the spare parts, specialized tools, and test equipment needed for common field repairs. These kits should be standardized across your organization, with clear inventory control and regular verification to ensure all components remain available when needed.

Preventive maintenance schedules must be implemented even during emergency operations. While the operational tempo may not allow for comprehensive maintenance, critical procedures for generators, cooling systems, and other high-wear components must continue on modified schedules to prevent catastrophic failures.

Remote diagnostic capabilities enable technical specialists to monitor equipment health and identify potential problems before they cause failures. Modern emergency communications equipment includes comprehensive telemetry systems that transmit operational parameters over available backhaul connections, allowing centralized monitoring of deployed assets.

Technical documentation for all deployed equipment must be readily available to field technicians, preferably in both digital and physical formats. These materials should include not just manufacturer documentation, but also field-expedient repair procedures developed based on previous emergency experiences.

Field Observations: Resource Sharing in Action

One of the most impressive aspects of the broader Hurricane Otis response was the collaboration between different telecommunications providers, government agencies, and humanitarian organizations to overcome resource limitations. I witnessed several examples of resource sharing that significantly enhanced the overall response effectiveness.

Major carriers shared transportation assets, with vehicles from one company carrying equipment from multiple providers to maximize delivery efficiency in an environment with limited access routes. This coordinated approach reduced redundant transportation efforts and ensured critical equipment reached priority locations regardless of which organization owned it.

I observed how telecommunications providers integrated their power generation resources, establishing shared generator sites that supported equipment from multiple companies. This approach optimized fuel usage and simplified the logistics of power maintenance during extended operations.

Perhaps most significantly, technical personnel from different organizations worked together to implement complementary solutions. Satellite providers established backhaul connections that cellular providers then utilized for their temporary base stations, while government agencies provided security and access permissions that benefited all technical teams.

These collaborations weren't entirely spontaneous—they built upon pre-existing relationships and agreements established before the disaster. The most effective resource sharing occurred between organizations that had already developed mutual aid frameworks, conducted joint training, and established clear protocols for emergency cooperation.

Case Study: Equipment Success Stories

The response to Hurricane Otis demonstrated both the value of comprehensive equipment planning and the importance of adaptability when facing unprecedented challenges. Several equipment deployments proved particularly noteworthy for their impact on the overall emergency response.

My Experience: First Critical Connection

The most impactful moment of our Hughes response to Hurricane Otis came with the establishment of the first reliable communications link at the Mexican Red Cross command post. This single VSAT connection broke the information blackout that had descended over the region, providing the first accurate assessment of the disaster's true magnitude to government authorities and relief organizations.

We faced significant challenges during this installation. Road conditions forced us to transport the final portions of equipment by foot. The installation site lacked proper infrastructure for mounting, requiring improvised solutions using materials available on-site. Despite these obstacles, the team completed the installation and established connectivity within hours of arriving at the location.

This connection enabled the coordination of search and rescue operations, facilitated the distribution of emergency supplies, and allowed responders to report critical needs back to command centers in Mexico City. It wasn't a sophisticated network by normal standards, but in that moment, it became the critical information lifeline for the entire emergency response.

Field Observations: Equipment Prioritization

Throughout the broader Hurricane Otis response, I observed how the decision to prioritize equipment for emergency services coordination rather than attempting to immediately restore public communications proved crucial to the overall effectiveness of the response. By establishing a network of communications links connecting emergency operations centers, hospitals, and government coordination facilities, telecommunications providers created a communications backbone that enabled the coordination of all other response activities.

The deployment of portable cellular infrastructure, including both COWs and smaller portable base stations, significantly reduced public anxiety by restoring familiar communications capabilities. By the third day of operations, cellular providers had established multiple cellular nodes providing service to a substantial portion of Acapulco's population.

Perhaps the most significant lesson from the equipment perspective came in the area of power management. Despite comprehensive planning, the extended duration of commercial power outages and challenges in fuel delivery created serious constraints on operations. The hybrid power systems deployed by some organizations, combining generators with solar arrays and advanced battery management, proved crucial to maintaining operations at sites where fuel deliveries were delayed by road conditions.

These experiences from Hurricane Otis highlight that effective telecommunications emergency response requires not just having appropriate equipment, but having it organized, deployed, and supported according to clear priorities that align with the broader emergency response effort. The organizations that demonstrated this capability made a profound difference in the overall disaster response and the affected communities' recovery.

CHAPTER 4: RESPONSE PROTOCOLS AND PROCEDURES

When emergency strikes and telecommunications infrastructure fail, having clear protocols and procedures can make the difference between an effective response and a chaotic one. This chapter explores the structured approach to telecommunications emergency response, from initial activation through coordinated deployment and operation.

The Activation Cascade

The emergency response begins well before teams arrive in the disaster zone. A systematic activation process ensures that the right resources are mobilized quickly and deployed effectively when communications fail.

The activation cascade should follow a clear sequence that begins with early warning monitoring and continues through a structured escalation process. Organizations with effective emergency response capabilities maintain continuous monitoring of potential threats—tracking weather systems, seismic activity, and other potential disaster triggers. This monitoring allows preliminary preparation even before impacts occur.

Pre-event preparation should begin as soon as a potential emergency is identified. This includes alerting response team members, verifying equipment readiness, and establishing initial communication with relevant government agencies. For predictable events like hurricanes, this preparation may begin days in advance, while sudden disasters like earthquakes will compress this timeline dramatically.

The formal activation decision should follow clear triggering criteria established in advance. These criteria might include specific impact thresholds (e.g., Category 3+ hurricane), formal requests from government agencies, or verified reports of significant infrastructure damage. Having predetermined activation criteria prevents delays caused by decision paralysis when information is limited or conflicting.

Once activated, the initial team deployment sequence should follow a standardized process. Typically, this begins with deploying assessment teams, followed by rapid response units carrying critical communications equipment. Command and control functions are established either remotely or on-site depending on the disaster's nature and scope. Throughout this process, clear communication about deployment status, arrival times, and initial findings must be maintained with all stakeholders.

Finally, the activation cascade includes escalation frameworks that determine when additional resources are required. These frameworks establish thresholds for expanding the response, whether through deploying additional teams, activating mutual aid agreements with other providers, or requesting government support. Effective escalation processes prevent both under-response to major disasters and over-commitment of resources to manageable situations.

The CLEAR Assessment Model

Effective response begins with accurate assessment. The CLEAR model provides a structured framework for evaluating disaster impacts on telecommunications infrastructure and determining appropriate response measures.

C - Communications Infrastructure Status

The first assessment priority is determining what communications infrastructure remains functional, what has been damaged, and what has been completely destroyed. This includes evaluating:

- Cell tower structural integrity and functionality
- Fiber optic network continuity and damage points
- Satellite earth station status
- Microwave link alignment and operation
- Power supply to critical infrastructure
- Data centers and network operation centers

This assessment must extend beyond your own infrastructure to include other carriers and service providers, as the overall communications ecosystem will determine both restoration priorities and technical approach.

L - Logistical Challenges and Access

Once infrastructure status is known, teams must assess the logistical environment that will determine how restoration efforts can proceed. This includes evaluating:

- Transportation network status (roads, bridges, airports, ports)

- Physical access to key infrastructure sites
- Security conditions in affected areas
- Available local resources (fuel, power, water, accommodations)
- Weather conditions and forecasts that might affect operations
- Geographical challenges specific to the disaster zone

Effective logistical assessment prevents situations where teams or equipment are deployed to areas they cannot safely or effectively reach.

E - Environmental Hazards and Conditions

Disaster zones often contain numerous environmental hazards that threaten both personnel safety and equipment function. Assessment must identify:

- Structural hazards from damaged buildings
- Chemical, biological, or radiological contamination
- Flooding and water hazards
- Fire risks
- Extreme weather conditions
- Disease vectors and health hazards
- Geological instability (landslides, sinkholes, aftershocks)

This environmental assessment ensures that emergency response doesn't create additional risks or place personnel in unreasonable danger.

A - Alternative Solutions Inventory

Based on the assessments above, teams must quickly inventory available solutions that match the specific disaster conditions. This includes:

- Deployable communication assets available for immediate use
- Backup systems and their status
- Alternative routing possibilities through surviving infrastructure
- Mutual aid resources available from other providers
- Government and military communications assets
- Novel or improvised solutions specific to the situation

This inventory should identify not just what technologies are theoretically available, but which can be practically deployed given the specific conditions identified in earlier assessments.

R - Resource Requirements Analysis

The final assessment component determines what resources will be needed to implement the identified solutions. This includes:

- Personnel requirements (technical specialties, number of staff, deployment duration)
- Equipment needs (communications hardware, power generation, support systems)
- Logistical support (transportation, fuel, water, food, shelter)
- Security requirements (escorts, site security, equipment protection)

- External coordination needs (government agencies, other providers, international organizations)

This comprehensive resource analysis ensures that the response can be properly scaled and sustained throughout the emergency period.

The CLEAR assessment model should be applied systematically but rapidly, with initial assessments completed within hours of disaster impact when possible. The model is iterative, with assessments refined as more information becomes available and conditions evolve. Most importantly, the assessment must translate directly into action, with each component informing specific aspects of the response plan.

My Experience: Rapid Assessment During Hurricane Otis

During Hurricane Otis, my role involved coordinating the rapid assessment process from our headquarters while our installation teams provided critical on-the-ground information. The unprecedented intensification of the hurricane from Category 1 to Category 5 in less than 24 hours compressed our normal assessment timeline dramatically.

With communications already failing across Acapulco, we had to rely on a combination of official reports, weather data, satellite imagery, and the limited information coming from emergency services radios to build our initial assessment. This highlighted the importance of having multiple information sources when normal assessment channels are compromised.

Our teams reported conditions far worse than initial official assessments had indicated. The infrastructure damage was nearly total in some areas, with cellular towers collapsed, fiber lines severed, and power completely unavailable. These firsthand reports from our installation personnel proved invaluable in understanding the true extent of the disaster and the specific challenges we would face in restoring communications.

The most critical insight from our assessment process came in identifying the Mexican Red Cross as the priority for initial communications restoration. Their central role in coordinating emergency response meant that establishing connectivity for their command post would have a multiplier effect, enabling numerous other life-saving activities. This prioritization decision, made during our rapid assessment phase, guided our entire response effort.

Prioritization Frameworks

When disaster strikes, the demand for communications restoration inevitably exceeds available resources. Effective emergency response requires clear frameworks for determining which needs will be addressed first, which must wait, and sometimes, which cannot be met during the initial emergency phase.

Strategic prioritization should follow established hierarchies that place life-saving functions first, operational coordination second, and broader public communications third. This hierarchy recognizes that while all communications have value, some are directly linked to saving lives in the immediate aftermath of disaster.

Tactical prioritization involves applying these strategic principles to specific decisions about where to deploy limited technical resources. This requires balancing multiple factors including:

- Humanitarian impact (how directly the communications will support life-saving operations)

- Technical feasibility (how quickly and reliably service can be restored)
- Resource requirements (how much equipment and personnel will be needed)
- Sustainability (how long the solution can be maintained in field conditions)
- Geographic coverage (how many people or critical functions will be served)

The most effective prioritization frameworks combine clear principles with sufficient flexibility to address unique disaster conditions. They should be established before disasters occur, understood by all team members, and communicated clearly to external stakeholders to manage expectations and prevent conflict over resource allocation.

Field-Proven Protocols: Priority Connectivity Points

Based on our disaster response experience, telecommunications teams should establish connectivity in this order of priority:

1. Emergency service headquarters: Red Cross, Fire Department, Police, Municipal/State Medical Services, National Guard
2. Local and regional hospitals
3. Command post base camps
4. Shelters
5. Public squares that can serve as community connectivity hubs

This prioritization ensures that telecommunications first enable coordinated emergency response, then support community recovery and family reconnection.

Coordination Protocols

Effective telecommunications emergency response never happens in isolation. It requires coordinated effort with numerous other entities including government agencies, military units, other service providers, and humanitarian organizations. Clear coordination protocols are essential for ensuring that these diverse stakeholders work together effectively during crisis conditions.

Government agency integration represents one of the most critical coordination requirements. This includes working with emergency management agencies that typically coordinate overall disaster response, telecommunications regulatory authorities that may need to grant emergency authorizations, and local governments that control access to many areas and facilities. Effective coordination with these agencies requires established relationships before disasters occur, designated liaison personnel during emergencies, and clear communication channels that function even when normal infrastructure fails.

Industry collaboration becomes particularly important during major disasters when no single provider has sufficient resources to address all needs. Telecommunications companies that normally compete must work together to restore critical services, often sharing resources, information, and even personnel. This collaboration should be guided by pre-established mutual aid agreements that address resource sharing, cost reimbursement, liability concerns, and operational coordination.

International coordination may be necessary during catastrophic events that require resources beyond what's available within the affected country. This coordination involves complex considerations including customs clearance for emergency equipment, frequency authorizations for communications devices, and permissions for foreign technical personnel to operate in the disaster zone. Organizations with international response capabilities should establish the necessary diplomatic relationships and documentation templates before emergencies occur.

Coordination mechanisms should include:

- Regular coordination meetings (virtual or in-person) with all key stakeholders
- Shared situational awareness through standardized reporting formats
- Clear deconfliction procedures to prevent duplication or interference
- Unified command structures that integrate telecommunications with broader emergency response
- Joint planning processes for resource allocation and prioritization

These coordination mechanisms ensure that limited telecommunications resources achieve maximum impact within the overall emergency response effort.

Documentation Requirements

While documentation may seem secondary during emergency operations, thorough record-keeping serves several essential functions in telecommunications disaster response.

Operational documentation provides the foundation for effective coordination, capturing decisions, actions, resource deployments, and technical configurations. This documentation ensures continuity when personnel change, prevents duplication of efforts, and enables accurate situation reports to leadership and external stakeholders. During extended operations, comprehensive documentation becomes increasingly important as the response evolves, and initial responders are replaced by fresh teams.

Technical documentation captures the specific details of emergency installations including equipment configurations, network settings, power requirements, and site information. This documentation is essential for maintaining and eventually transitioning these temporary solutions. It also provides valuable reference for future emergency deployments, capturing both standard approaches and improvised solutions developed under field conditions.

Financial documentation tracks all costs associated with the emergency response including equipment, personnel time, transportation, supplies, and external services. This documentation is essential for cost recovery through insurance, government reimbursement programs, or regulatory mechanisms. Without proper financial documentation, organizations may absorb significant unrecovered costs from their emergency response efforts.

Legal and regulatory documentation addresses the special permissions, waivers, and authorizations often granted during emergencies. This might include temporary frequency allocations, expedited permits for installations, or special access authorizations to restricted areas. Properly documenting these temporary arrangements is essential for both compliance and eventual transition back to normal operations.

Effective documentation systems should balance comprehensiveness with operational practicality. They must capture essential information without creating administrative burdens that impede emergency response. Modern approaches often use digital tools like tablets with specialized forms, photographic documentation, and GPS-tagged records to streamline the documentation process while maintaining necessary detail.

Security Considerations in Disaster Zones

My Experience: New Security Challenges During Hurricane Otis

Hurricane Otis represented a watershed moment in my telecommunications emergency response experience. For the first time in my years working in the telecommunications industry and maintaining close relationships with emergency organizations, I witnessed how security concerns—primarily related to criminal organizations—actively interfered with aid reaching its destination.

While I personally monitored the situation remotely on a 24/7 basis, it was our Hughes installation teams who were the true heroes, experiencing these unprecedented security challenges firsthand. From my coordination position, I maintained constant communication with our teams as they worked to establish those critical initial satellite connections for the Mexican Red Cross.

Our installers reported operating in an environment where the normal social fabric had been compromised. The breakdown of local infrastructure, combined with desperation among some of the population and the opportunistic activities of criminal elements, created dangerous conditions unlike anything we had encountered in previous disaster responses.

We were forced to implement extreme security measures for our technical personnel. Every movement required careful planning, coordination with security forces, and constant situational awareness. Equipment that would normally attract attention primarily for its technical value now represented a potential target for theft or exploitation, requiring additional protective measures during transport and installation.

This experience fundamentally changed my understanding of telecommunications emergency response requirements. Through the firsthand accounts of our installation teams and my continuous monitoring of the situation, I realized that technical expertise and appropriate equipment are necessary but insufficient conditions for successful deployment. Without adequate security provisions, even the most well-prepared technical teams cannot fulfill their mission of restoring critical communications.

Field Observations: Security Adaptations in the Broader Response

Throughout the Hurricane Otis response, I observed how larger telecommunications providers adapted their operations to address these security challenges. Major carriers deployed their technical teams with security escorts, often coordinating directly with military or specialized police units to ensure safe access to installation sites.

Security considerations began to influence technical decisions in ways I had never previously observed. Site selection for temporary communications equipment now included security vulnerability assessments alongside the traditional technical criteria. Some potentially optimal locations from a coverage perspective were rejected due to security concerns, with deployments redirected to areas that offered better protection for both equipment and personnel.

Communications about deployment activities became much more restricted, with information about equipment movements, installation schedules, and team locations shared only on a need-to-know basis. This operational security approach, more typical of military operations than civilian telecommunications work, became essential to protecting teams and resources.

Perhaps most significantly, I noticed how telecommunications providers began to integrate security personnel directly into their emergency response teams rather than treating security as an external service. These embedded security specialists participated in planning meetings, contributed to risk assessments, and helped develop operational protocols specifically designed to address the unique threats present in the Hurricane Otis environment.

Developing a Security-Conscious Response Framework

Based on these experiences and observations, telecommunications organizations developing emergency response capabilities must now incorporate comprehensive security planning into their preparations. This includes:

Threat Assessment Protocols: Developing methodologies to quickly evaluate security conditions in disaster areas, identifying specific threats to telecommunications teams and equipment. These assessments must go beyond general situational awareness to include specific vulnerability analysis for technical operations.

Security Integration in Team Structure: Including security specialists as core members of emergency response teams rather than treating security as an external consideration. These personnel should have both security expertise and sufficient understanding of telecommunications operations to provide contextually appropriate protection.

Secure Communications Practices: Implementing operational security measures for all communications related to team movements, equipment transportation, and installation schedules. This includes both technical measures to protect communication channels and procedural controls on information sharing.

Coordination with Official Security Forces: Establishing formal relationships with military, law enforcement, and other security agencies before disasters occur. These relationships should include clear protocols for requesting security support, communication channels for coordination during operations, and joint training when possible.

Personnel Security Training: Providing all team members with security awareness training that addresses the specific risks encountered in disaster environments. This training should cover situational awareness, threat recognition, de-escalation techniques, and emergency response procedures for security incidents.

Secure Equipment Design: Considering security implications in the selection and configuration of emergency communications equipment. This might include features like remote disabling capabilities, tracking systems, tamper-evident enclosures, and designs that minimize attractive nuisance factors.

The security dimension of telecommunications emergency response has evolved from a peripheral consideration to a central determining factor in operational success. Organizations that fail to adapt to this new reality risk not only equipment loss but, more importantly, personnel safety. As we witnessed during Hurricane Otis, even the most urgent humanitarian communications missions cannot proceed without adequate security measures in place.

This experience has profoundly influenced my approach to telecommunications emergency response planning. I now consider security planning to be as fundamental as technical preparation, and I strongly encourage all telecommunications organizations to integrate comprehensive security measures into their emergency response frameworks.

Field-Proven Protocols: Operational Safety Guidelines

The following activities are strictly prohibited during disaster response:

- Working in excessive wind, rain, or electrical storms
- Traveling in areas where authorities have not confirmed minimum safety conditions
- Driving at night in affected areas
- Working in isolation—always work alongside another company team and ideally with official security vehicle escort
- Departing from or returning to base without notification—the company must know your location at all times

Field-Proven Protocols: Basic Disaster Response Principles

When responding to emergencies, remember these fundamentals:

- Personal safety comes first—if you're not okay, you cannot help others
- Become agents of assistance, not part of the emergency
- If your vehicle isn't in proper condition, don't deploy
- Plan for an extended deployment of days or even weeks—assume there will be no food services, accommodations, or fuel available

- Establish clear routes for arrival and departure back to your base point
- Return to basics: use maps and compass for navigation when digital systems are unavailable

Communication Strategy

Beyond restoring communications infrastructure, telecommunications emergency teams must implement their own effective communication strategy. This strategy serves multiple purposes: coordinating internal operations, managing relationships with external stakeholders, and supporting the information needs of affected communities.

Internal communications ensure that all members of the emergency response team maintain situational awareness and operational coordination. This includes regular status updates, shift handovers, safety briefings, and technical coordination. Even when deploying advanced communications technology for others, emergency teams often need to maintain their own separate communications channels using satellite phones, emergency radios, or other systems that function independently of the infrastructure being restored.

External stakeholder communications maintain relationships with government agencies, other service providers, humanitarian organizations, and corporate leadership. These communications typically focus on operational coordination, resource requests, regulatory compliance, and progress reporting. The format and frequency of these communications should be standardized to ensure consistency and reduce the reporting burden on operational teams.

Public information communications address the information needs of affected communities regarding service restoration. During disasters, uncertainty about when communications will be restored creates significant anxiety among affected populations. Clear, honest communications about restoration priorities, progress, and timelines can reduce this anxiety even when full restoration will take time. These communications must balance transparency with security considerations, particularly in environments where telecommunications equipment might be targeted.

Media relations communications require particular attention during high-profile disasters. Media organizations will seek information about the telecommunications situation, restoration efforts, and impacts on affected communities. Organizations should designate specific spokespersons, develop clear messaging guidelines, and establish approval processes for media communications to ensure accuracy and consistency while preventing operational disruption.

The most effective communication strategies encompass all these dimensions while recognizing the limited bandwidth—both literal and figurative—available during emergency operations. They prioritize operational effectiveness while maintaining necessary transparency with all stakeholders.

CHAPTER 5: TRAINING AND SIMULATION PROGRAM

The effectiveness of telecommunications emergency response depends largely on preparation that occurs long before disaster strikes. This chapter explores how to develop comprehensive training and simulation programs that build both individual capabilities and team effectiveness for disaster conditions.

Skills Development Curriculum

Technical emergency skills form the foundation of telecommunications disaster response. These skills go beyond normal technical capabilities to include specific competencies needed in emergency conditions. Technical training should emphasize:

Improvised repairs using available materials when standard components aren't available. This includes creative approaches to antenna mounting, alternative power connections, and field-expedient weatherproofing. Technical staff should practice creating functional solutions with limited resources to develop the ingenuity needed during actual disasters.

Degraded mode operations for all key systems, ensuring technicians can configure and maintain equipment under suboptimal conditions. This includes operating with reduced power, limited bandwidth, compromised environmental conditions, and partial system functionality. Teams should be comfortable with manual procedures for systems that normally operate automatically, ensuring capability even when advanced features are unavailable.

Alternative solution engineering that adapts technologies for unintended uses when circumstances require. This might include repurposing commercial equipment for emergency applications, creating hybrid solutions that combine components from different systems, or developing temporary interfaces between incompatible systems. This creative engineering capability becomes particularly valuable when standard solutions cannot be implemented due to disaster conditions.

Field maintenance under adverse conditions, including troubleshooting and repairs in harsh environments without standard facilities or tools. Technicians should be able to diagnose problems with limited test equipment, perform repairs with basic tools, and maintain system functionality despite environmental challenges like extreme temperatures, moisture, dust, or unstable power.

Operational emergency capabilities extend beyond technical skills to include the practical aspects of functioning in disaster environments. This training should cover:

Rapid deployment procedures for all emergency equipment, emphasizing speed, efficiency, and adaptation to field conditions. Teams should regularly practice deployment under time pressure and in challenging environments, developing the muscle memory and procedural familiarity that enables quick setup when every minute counts.

Field assessment techniques for evaluating infrastructure damage, site conditions, and operational requirements with limited information and resources. This includes both technical assessment of telecommunications systems and broader evaluation of environmental conditions, security situations, and logistical challenges that affect operations.

Crisis decision-making frameworks that enable sound judgments under pressure with incomplete information. Operational staff should practice structured approaches to priority setting, resource allocation, and risk assessment that can be applied consistently even in chaotic conditions. This training should emphasize both analytical tools and recognition of cognitive biases that can affect decision quality during high-stress situations.

Resource management under constraints, including techniques for maximizing effectiveness with limited supplies, equipment, and personnel. This includes prioritization frameworks, conservation strategies, and creative approaches to resource limitations that inevitably occur during extended emergency operations.

Emergency leadership development addresses the human dimensions of disaster response. This training focuses on:

Crisis communication techniques for maintaining clear, effective information flow during emergency conditions. Leaders should develop skills in concise messaging, active listening under pressure, and maintaining communication discipline across distributed teams operating in challenging environments.

Team management in high-stress environments, including maintaining morale, managing fatigue, addressing conflicts, and sustaining performance during extended operations. Leaders should understand the psychological impacts of disaster work and develop specific techniques for supporting team effectiveness under these conditions.

Ethical decision-making under pressure, particularly regarding resource allocation, risk acceptance, and prioritization choices that affect communities and individuals. Leaders should practice applying ethical frameworks to the complex, often no-win scenarios that emerge during major disasters when needs inevitably exceed available resources.

Psychological support for team members experiencing the physical and emotional challenges of disaster response. Leaders need skills in psychological first aid, stress management, trauma recognition, and appropriate referral for more serious issues. This training should emphasize both supporting others and recognizing one's own psychological needs during emergency operations.

Simulation Exercises

Effective telecommunications emergency response requires more than individual skills—it demands seamless team performance under extreme conditions. Simulation exercises provide the essential practice environment for building this capability.

Tabletop scenarios offer a controlled environment for practicing decision-making, coordination, and problem-solving without the logistical challenges of field deployment. These exercises typically gather team members around a simulated emergency scenario, with participants responding to evolving conditions presented by facilitators. Effective tabletop exercises:

- Present realistic, detailed scenarios based on plausible disaster conditions
- Introduce unexpected complications and cascading effects
- Require actual decision-making rather than theoretical discussion
- Include representation from all functional areas and external partners
- Provide immediate feedback and learning opportunities

Such exercises are particularly valuable for practicing coordination procedures, testing decision frameworks, and identifying gaps in existing plans or capabilities.

Decision-making exercises focus specifically on the critical choices that emergency responders must make under pressure. These simulations typically present participants with complex problems requiring immediate action despite incomplete information, competing priorities, and limited resources. Effective decision exercises:

- Impose realistic time constraints that create genuine pressure
- Present information in the imperfect, fragmentary way it would arrive during actual emergencies
- Require balancing multiple objectives with no perfect solutions
- Include ethical dimensions and human impact considerations
- Provide structured debriefing to examine decision quality and process

These exercises develop both individual judgment and team alignment around decision frameworks that can be applied during actual emergencies.

Resource allocation simulations address the perpetual challenge of managing limited resources during disaster response. These exercises require participants to prioritize needs, allocate scarce equipment and personnel, and adjust resource distribution as conditions change. Effective resource allocation simulations:

- Present realistic inventory constraints based on actual capabilities
- Require sequential allocation decisions with downstream consequences
- Include unexpected resource losses and requirement changes
- Impose logistical limitations on resource movement
- Incorporate multiple competing demands from various stakeholders

Such exercises help teams develop systematic approaches to resource management that can be applied under the pressure of actual emergencies.

Field training operations move beyond tabletop exercises to physical deployment of equipment and personnel in realistic conditions. These exercises test both technical skills and operational capabilities in environments that approximate actual disaster conditions. Effective field training:

- Deploys actual emergency equipment using standard procedures
- Operates in challenging environments (remote locations, difficult terrain, adverse weather)
- Includes realistic logistical and communication constraints
- Requires complete setup and operation of systems, not just symbolic actions
- Incorporates unexpected problems requiring field solutions

Field exercises identify practical challenges that don't emerge in tabletop simulations, while building team confidence and cohesion through shared accomplishment.

Full-scale disaster simulations represent the most comprehensive training approach, integrating all aspects of emergency response in scenarios that closely mimic actual disasters. These complex exercises often involve multiple organizations and incorporate all dimensions of disaster response from initial activation through extended operations. Effective full-scale simulations:

- Create immersive scenarios with realistic environmental conditions
- Involve actual emergency equipment deployed according to standard protocols

- Include participation from external partners and government agencies
- Incorporate realistic constraints on communications, transportation, and resources
- Extend beyond initial deployment to include sustained operations and rotation planning

These exercises provide the most realistic preparation for actual disaster response, though their complexity and cost typically limit them to annual or semi-annual frequency.

After-action review processes transform simulation experiences into organizational learning. These structured assessments examine what worked, what didn't, and why, creating specific improvement plans for both individuals and teams. Effective after-action reviews:

- Maintain psychological safety to encourage honest assessment
- Focus on systemic factors rather than individual blame
- Identify both strengths to maintain and weaknesses to address
- Create specific, actionable improvement plans with clear ownership
- Document lessons learned for future reference and training

These reviews close the learning loop, ensuring that each simulation contributes to ongoing capability development rather than merely testing current performance.

Performance Measurement

Effective training requires objective assessment of both individual and team performance. A comprehensive measurement system helps identify development needs, track improvement, and verify readiness for actual emergency deployment.

Response time metrics measure how quickly teams can activate, deploy, and establish emergency communications in various scenarios. These metrics typically include:

- Activation to deployment: Time from emergency notification to team departure
- Deployment to arrival: Transit time to affected location
- Arrival to first solution: Time to establish initial communications capability
- Coverage restoration milestones: Progressive expansion of service over time

These metrics should be tracked consistently across training exercises, with targets based on operational requirements and continuous improvement goals.

Effectiveness indicators measure the quality and impact of emergency communications solutions beyond mere speed of deployment. These metrics might include:

- Population reconnected: Number of people regaining communications access
- Critical services restored: Essential functions supported by emergency communications
- Bandwidth delivered: Capacity provided for various service types

- Service reliability: Stability and consistency of emergency communications
- Resource efficiency: Results achieved relative to resources deployed

These indicators help teams focus on impact rather than just activity, ensuring that emergency response delivers meaningful results for affected communities.

Capability development tracking measures how individual and team capabilities evolve over time through training and experience. These metrics might include:

- Technical proficiency assessments for core emergency skills
- Decision quality evaluation in simulation exercises
- Team coordination effectiveness during complex scenarios
- Leadership performance under various emergency conditions
- Problem-solving adaptability when facing unexpected challenges

This tracking helps identify both individual development needs and team capability gaps that require additional training or resources.

Comprehensive performance measurement systems incorporate all these dimensions while recognizing the unique nature of emergency response. Unlike many business metrics, emergency response measurement must account for the inherent variability of disaster conditions while still providing meaningful comparisons across exercises and actual deployments.

The most effective measurement approaches combine quantitative metrics with qualitative assessment, recognizing that numbers alone cannot capture all aspects of emergency response effectiveness. They also involve multiple perspectives—including team members, leaders, and external partners—to ensure a complete picture of performance and development needs.

My Experience: Training Value Revealed During Hurricane Otis

While monitoring our Hughes team's response to Hurricane Otis, I witnessed firsthand how previous training directly influenced operational effectiveness in the field. Our installation teams drew upon their simulation experience when faced with the challenging conditions in Acapulco, applying practiced skills to solve problems that might otherwise have prevented successful deployment.

The most valuable aspect of our previous training proved to be the emphasis on improvisation with limited resources. When standard mounting equipment proved unsuitable for the damaged structures available at the Red Cross command post, our technicians quickly developed alternative mounting solutions using materials they found on site. This ability to adapt standard procedures to unexpected conditions—practiced repeatedly during training—enabled critical communications restoration despite challenging circumstances.

Another training element that proved particularly valuable was the practice of rapid but thorough site assessment. Our teams had developed a structured approach to evaluating potential installation locations, considering not just technical factors but also physical security, environmental exposure, and accessibility for maintenance. This systematic assessment process, refined through multiple simulation exercises, guided effective site selection under the chaotic conditions following Hurricane Otis.

Perhaps most importantly, our teams demonstrated the decision-making confidence that comes from regular training scenarios. When faced with unexpected challenges and limited time, they didn't hesitate or second-guess their approach—they applied the decision frameworks practiced in training to make sound choices quickly and proceed with implementation. This decisiveness saved precious hours during the critical initial response period.

The Hurricane Otis experience reinforced my belief that training investment delivers its greatest return precisely when conditions are at their worst. The skills and mindsets developed through realistic simulation became not just useful but essential when our teams faced the real-world challenges of this unprecedented disaster.

CHAPTER 6: PSYCHOLOGICAL ASPECTS OF EMERGENCY RESPONSE

The effectiveness of telecommunications emergency response depends not only on technical capabilities and equipment but also on the human dimension. This chapter explores the psychological aspects of disaster response, addressing how to prepare, support, and sustain the mental wellbeing of team members operating under extreme conditions.

Mental Preparation and Resilience

Disaster environments create psychological challenges that can impair performance and lead to both short-term operational issues and long-term personal consequences. Proper mental preparation helps teams respond effectively while protecting their psychological wellbeing.

Pre-deployment psychological briefing provides essential context and preparation before team members enter disaster environments. These briefings should include realistic previews of conditions they may encounter, normal stress responses they might experience, and coping strategies for managing psychological pressure. Effective briefings neither minimize challenges nor catastrophize them, instead providing honest information that helps team members develop realistic expectations and preparation strategies.

The briefing should address the potential emotional impact of witnessing disaster effects, including destroyed infrastructure, community disruption, and in severe cases, human suffering.

While telecommunications responders aren't typically first-line medical responders, they still operate in environments with significant trauma and must be prepared for the emotional impact this can have.

Stress management techniques should be taught and practiced before deployment rather than introduced during crisis. These techniques include tactical breathing for immediate stress reduction, cognitive reframing to maintain perspective, mindfulness practices for emotional regulation, and structured routines that provide stability amid chaos. Team members should practice these techniques regularly, so they become automatic responses during high-stress situations.

Decision-making under extreme pressure requires both psychological preparation and practical frameworks. Team members should understand how stress affects cognitive function, including common effects like narrowed attention, reduced working memory, and altered risk perception. Training should include specific strategies for maintaining effective decision quality despite these effects, including structured decision processes, team decision approaches, and techniques for recognizing when stress is impairing judgment.

Sleep deprivation management becomes increasingly important during extended emergency operations. Teams should be educated about the cognitive and emotional effects of sleep loss, trained in sleep optimization techniques for challenging environments, and provided with clear protocols for mandatory rest periods. Leaders must understand that sleep is not a luxury but an operational necessity, as sleep-deprived personnel make more errors, process information more slowly, and have impaired judgment and emotion regulation.

Dealing with traumatic scenes and situations requires specific preparation, even for telecommunications responders who aren't directly involved in medical response or search and rescue. Team members should receive basic training in psychological first aid, including how to process their own reactions to distressing situations and how to support colleagues who may be struggling. This training should normalize emotional responses while providing practical strategies for maintaining functional capacity.

Team Psychological Support

Beyond individual preparation, the team environment plays a crucial role in sustaining psychological resilience during emergency operations.

Buddy systems and peer support structures formalize the natural human tendency to connect during challenging situations. These systems pair team members who monitor each other's wellbeing, provide emotional support, recognize signs of excessive stress, and intervene when necessary. Effective buddy systems create accountability for self-care while reducing the stigma around acknowledging stress responses. The buddy approach also creates a first line of psychological support that can address many issues before they require formal intervention.

Rotation and rest requirements must be clearly established and rigorously enforced. These requirements should specify maximum operational periods, minimum rest intervals, and mandatory down days during extended deployments. While operational pressures create temptation to extend shifts or reduce rest periods, leaders must recognize that this apparent short-term gain creates significant risks to both team health and mission effectiveness. Research consistently shows that well-rested teams accomplish more over extended operations than those working beyond sustainable limits.

On-site psychological first aid capability should be integrated into emergency response teams during significant deployments. This capability might range from team members with specialized training to professional mental health providers embedded with the response team, depending on the scale and intensity of the operation. On-site support allows for immediate intervention when team members experience acute stress reactions, providing early assistance that can prevent escalation to more serious conditions.

Post-deployment decompression represents a critical transition between the high-intensity disaster environment and return to normal operations. This process should include structured opportunities to process experiences, recognize accomplishments, address concerns, and prepare for reintegration into regular life. Effective decompression isn't merely a procedural requirement but a meaningful process that helps team members contextualize their experience and begin the transition back to normal routines.

Long-term follow-up and support acknowledge that the psychological impacts of disaster response can emerge weeks or months after deployment. Organizations should implement structured check-ins at regular intervals, provide access to professional support resources, and maintain peer connections that facilitate ongoing processing of the experience. This long-term approach recognizes that psychological impacts don't necessarily end when the deployment concludes and provides pathways for assistance if delayed reactions emerge.

Leadership Under Crisis

Leaders play a pivotal role in shaping the psychological environment during emergency operations, with their behavior and decisions significantly influencing team resilience and effectiveness.

Maintaining team morale during extended emergency operations requires conscious attention and specific leadership practices. Effective approaches include recognizing accomplishments even amid challenges, maintaining transparent communication about the situation and response efforts, creating opportunities for team cohesion through shared meals or brief gatherings, and modeling resilient attitudes while acknowledging difficulties. Leaders who balance realistic acknowledgment of challenges with confidence in the team's capabilities create an environment where members can sustain motivation despite difficult conditions.

Ethical decision-making frameworks become particularly important during crisis when leaders face difficult choices with significant human impacts. Leaders should have established frameworks for navigating these decisions, including clear principles that guide resource allocation, risk acceptance, and prioritization. These frameworks help leaders make consistent decisions aligned with organizational values while reducing the psychological burden of making difficult choices in isolation.

Balancing urgency with staff wellbeing represents one of the most challenging leadership responsibilities during emergency response. Leaders must maintain the pace and focus needed for effective response while ensuring sustainable operation that doesn't compromise team health. This balance requires constant attention to signs of team stress, willingness to adjust operations when needed, and the courage to enforce rest and recovery despite operational pressures.

Recognition and acknowledgment during operations provide essential psychological support that sustains team effectiveness. Leaders should implement systematic approaches to recognizing contributions, celebrating achievements (even small ones), and acknowledging the challenges teams are facing. This recognition isn't merely ceremonial but serves a functional purpose in maintaining team motivation and cohesion during difficult conditions.

Managing conflicting priorities and demands requires both clear decision frameworks and effective communication. Leaders inevitably face competing requests from various stakeholders during emergency operations, creating potential for team confusion or misdirected efforts. Effective leaders establish clear prioritization criteria, communicate decisions transparently, and shield their teams from the cross-pressures that might otherwise create confusion or unnecessary stress.

Field Observations: The Human Factor in Disaster Response

Throughout multiple disaster responses, I've observed how the psychological dimension often determines the difference between successful operations and those that falter despite adequate technical capabilities and equipment.

The most effective response teams maintain a distinct psychological climate characterized by several key elements. They demonstrate "focused calm" – maintaining urgency without panic, addressing problems systematically rather than reactively. They practice "transparent optimism" – acknowledging challenges honestly while maintaining confidence in their collective capability to overcome them. They engage in "active caring" –

team members visibly looking out for each other's wellbeing without diminishing professional focus.

I've observed that teams with strong psychological preparation demonstrate significantly greater adaptability when facing unexpected challenges. Rather than becoming fixated on original plans when conditions change, they quickly adjust their approach, find creative alternatives, and maintain effectiveness despite setbacks. This adaptability stems not just from technical flexibility but from psychological resilience that allows them to process disappointment and refocus quickly.

Perhaps most significantly, I've noticed that the recovery trajectory after deployment varies dramatically based on how teams manage the psychological aspects during operations. Teams that implement good psychological practices – including adequate rest, processing of experiences, peer support, and leader attention to wellbeing – typically show faster recovery and fewer lingering effects than those that neglect these dimensions in favor of pure operational focus.

These observations reinforce the understanding that psychological factors aren't secondary considerations in emergency response but central determinants of both operational effectiveness and team wellbeing. Organizations that invest in this dimension find that it yields returns in both immediate performance and long-term capability.

CHAPTER 7: TECHNOLOGICAL INNOVATIONS IN EMERGENCY RESPONSE

The landscape of telecommunications emergency response continues to evolve as new technologies emerge and existing systems adapt to meet the challenges of disaster communications. This chapter explores current innovations and future directions that are transforming how we restore communications in crisis situations.

Emerging Technologies

Recent years have seen remarkable technological advances that are changing the possibilities for emergency telecommunications response. Understanding these emerging technologies helps organizations prepare for the next generation of disaster communications capabilities.

High-altitude platforms (HAPs) represent a promising middle ground between terrestrial and satellite systems for emergency communications. These platforms – including specialized aircraft, balloons, and unmanned aerial vehicles – operate in the stratosphere (typically 18-25 km altitude) and can provide coverage to areas up to 200 km in diameter. Their advantages for disaster response include rapid deployment compared to rebuilding ground infrastructure, wider coverage than terrestrial systems, lower latency than satellite communications, and flexibility in positioning to match specific disaster geographies.

Companies like Loon (formerly Google's Project Loon) demonstrated the potential of balloon-based HAPs for disaster recovery, most notably during Hurricane Maria in Puerto Rico. While some early commercial efforts have faced challenges, the technology continues to advance with both commercial and governmental applications under development.

Drone-based communications systems operate at lower altitudes than HAPs but offer even faster deployment and more precise positioning. Modern emergency response drones can carry various communications payloads including:

- Portable cellular base stations that create temporary coverage zones
- Wi-Fi access points for localized connectivity in critical areas
- Radio repeaters that extend emergency service communications
- Aerial survey equipment that rapidly maps infrastructure damage

These systems can be deployed within minutes of arriving at a disaster site, providing immediate communications while larger systems are being established. Advanced drone systems now feature extended flight times (some exceeding 24 hours with tethered power), automated flight patterns, and sophisticated payload management that allows for service optimization without continuous operator intervention.

Rapidly deployable low-earth orbit (LEO) solutions are transforming satellite communications for emergency response. Unlike traditional geostationary satellites, LEO constellations orbit at lower altitudes (typically 500-1,200 km), providing reduced latency, increased bandwidth, and coverage of polar regions. For disaster response, these systems offer:

- Higher bandwidth than traditional satellite solutions

- Smaller, more portable ground terminals
- Reduced latency that supports video conferencing and other interactive applications
- Coverage in areas traditionally challenging for satellite communications

While companies like Starlink (SpaceX), OneWeb, and Amazon's Project Kuiper are primarily developing these systems for commercial applications, their potential for disaster response is significant. Some providers have already demonstrated this capability, deploying services following natural disasters to support both emergency responders and affected communities.

AI-based network optimization represents another significant advancement for emergency communications. These systems use machine learning algorithms to:

- Automatically adjust network parameters based on changing conditions
- Predict usage patterns and preemptively allocate resources
- Identify network anomalies that might indicate developing problems
- Optimize service quality across limited bandwidth connections

In emergency situations where resources are constrained and conditions constantly change, these automated optimization capabilities can significantly improve service quality while reducing the management burden on technical teams already stretched thin by disaster conditions.

Augmented reality for field technicians has emerged as a powerful tool for telecommunications emergency response. These systems, typically using hardened tablets or specialized headsets, provide technicians with:

- Interactive installation guides overlaid on their visual field
- Remote expert assistance from specialists not on-site
- Visualization of underground infrastructure and hazards
- Real-time diagnostic information from equipment being serviced

These capabilities are particularly valuable in disaster environments where experienced technicians may be limited, conditions may be unfamiliar, and time pressure requires rapid, accurate work without extensive reference to manuals or repeated attempts.

Automated assessment systems using computer vision and sensor networks are transforming the initial phase of telecommunications emergency response. These systems incorporate:

- Drone-based optical and infrared imaging to assess infrastructure damage
- Machine learning algorithms that automatically identify damaged components
- LIDAR and photogrammetry that create detailed 3D models of affected areas
- Sensor networks that report the status of telecommunications infrastructure

These technologies enable faster, more comprehensive assessment of telecommunications damage following disasters, allowing for more effective prioritization and resource allocation from the earliest stages of response.

Data Management During Crisis

Modern telecommunications emergency response generates and depends upon vast amounts of data. Effective management of this information has become a critical capability for successful disaster response.

Critical network data backup systems ensure that essential configuration information, customer data, and operational parameters remain available even when primary systems are compromised. Modern approaches include:

- Geographically distributed backup repositories with automated synchronization
- Blockchain-based systems that ensure data integrity even when portions of the network fail
- Immutable storage technologies that prevent data corruption during disaster events
- Portable, hardened data repositories that can be physically transported to disaster zones

These systems guarantee that response teams have the information needed to restore services quickly, without depending on potentially compromised data centers in the affected area.

Cloud-based emergency operations centers represent a significant evolution from traditional fixed facilities. These virtual command environments offer several advantages for telecommunications response:

- Accessibility from any location with internet connectivity
- Resilience against local infrastructure failures

- Scalability to support varying team sizes and operational complexity
- Seamless information sharing across distributed teams
- Integrated communications, data visualization, and decision support tools

While physical command posts remain important for many aspects of emergency response, these cloud-based systems provide continuity of operations even when physical facilities are compromised or inaccessible.

Remote diagnostic capabilities have transformed how technical issues are identified and resolved during emergency operations. Modern systems feature:

- IoT sensors that continuously monitor equipment health
- AI-based anomaly detection that identifies developing problems
- Remote access to equipment configuration and performance data
- Predictive analytics that forecast potential failures before they occur

These capabilities allow scarce technical resources to be deployed more effectively, with remote experts handling diagnostic tasks while field personnel focus on physical installation and repair activities that require on-site presence.

Real-time coverage and usage analytics provide unprecedented visibility into the performance and impact of emergency communications. These systems deliver:

- Detailed maps showing service availability across the affected area

- Usage patterns that help identify underserved populations
- Performance metrics that highlight areas needing technical attention
- Impact assessments that demonstrate the value of restored services

This information allows for continuous optimization of the emergency response, with resources directed where they will have the greatest impact rather than following predetermined plans that may not match evolving conditions.

Predictive restoration modeling uses advanced analytics to forecast how different response approaches will affect service restoration. These systems incorporate:

- Historical data from previous disaster responses
- Current information about infrastructure damage
- Resource availability and deployment timelines
- Environmental factors affecting restoration efforts

The resulting models help leadership teams evaluate alternative approaches, make informed decisions about resource allocation, and provide realistic estimates of restoration timelines to external stakeholders.

Resource optimization algorithms automatically analyze complex emergency situations and recommend resource allocations that maximize impact. These systems can:

- Balance immediate tactical needs against longer-term strategic objectives
- Incorporate multiple constraints including team skills, equipment availability, and access limitations

- Adapt recommendations as conditions change
- Evaluate trade-offs between different objectives such as coverage area, bandwidth, and service reliability

While human judgment remains essential for final decisions, these algorithms help identify options that might not be apparent through traditional planning approaches and ensure that all relevant factors are considered despite the time pressure of emergency conditions.

Customer Solutions in Catastrophic Failures

Beyond restoring infrastructure, modern telecommunications emergency response increasingly includes specific solutions designed for end users affected by service disruptions.

Emergency customer communication systems ensure that service providers can maintain contact with affected populations even when normal communications channels fail. These systems typically include:

- Pre-disaster registration for emergency notifications
- Multi-channel delivery options including SMS, email, app notifications, and social media
- Location-based targeting to reach specific affected areas
- Message templates prepared for common emergency scenarios

These capabilities allow providers to deliver critical information about service status, restoration efforts, alternative communications options, and safety guidance when most needed.

Self-service disaster recovery solutions empower customers to restore basic communications without requiring direct technical support. These approaches include:

- Simplified instructions for reconfiguring devices after network changes
- Automated connectivity tools that find and connect to available networks
- Offline functionality that provides basic services even without network connectivity
- Peer-to-peer communication options that work without infrastructure

By enabling customers to establish basic connectivity themselves, these solutions reduce demand on scarce technical resources while providing more people with essential communications capabilities during the critical early phase of disaster response.

Community connectivity hubs provide centralized communications access in areas where widespread service restoration isn't immediately feasible. Modern implementations of this concept include:

- Portable, self-contained units with satellite backhaul, Wi-Fi distribution, and device charging
- Solar-powered designs that operate independently of local power infrastructure
- User management systems that ensure equitable access across affected populations
- Content caching that provides access to critical information even with limited backhaul bandwidth

These hubs serve both practical and psychological functions – providing essential communications capability while creating community gathering points that facilitate mutual support during crisis conditions.

Priority service frameworks ensure that limited communications resources are allocated to the most critical needs during partial service restoration. These systems include:

- Technical mechanisms that prioritize traffic from emergency services and critical infrastructure
- Access control systems that manage utilization of limited-capacity emergency networks
- Bandwidth management tools that ensure critical applications receive necessary resources
- Dynamic prioritization that adjusts based on evolving emergency conditions

While complete service restoration remains the ultimate goal, these frameworks ensure that limited interim capacity delivers maximum benefit to affected communities.

Alternative communication methods provided by telecommunications companies represent a growing aspect of emergency response. These approaches include:

- Offline messaging apps that store and forward communications when connectivity becomes available
- Low-bandwidth text-based services that function when data networks are compromised
- Mesh networking applications that create device-to-device communications without infrastructure

- Simplified voice services that require minimal bandwidth and infrastructure

By offering these alternatives during emergencies, providers help maintain essential communications capabilities even when full-service restoration takes time.

Customer education for emergencies has become increasingly important as communications systems grow more complex and users become more dependent on them. Effective education programs include:

- Pre-disaster guidance on preparing devices and accounts for emergency conditions
- Information about service limitations during different types of disasters
- Instructions for accessing emergency services when normal systems are compromised
- Recommendations for communications planning as part of overall emergency preparedness

This education helps set realistic expectations while empowering users to maintain essential communications during disaster conditions.

Innovation Case Study: Solutions Developed During Major Disasters

Major disasters have often served as catalysts for telecommunications innovation, with the urgent needs of affected communities driving creative solutions that later become standard components of emergency response.

Field Observations: Innovation During Hurricane Response

During major hurricane responses in recent years, I've observed several notable innovations that emerged from the necessity of addressing unprecedented challenges. These innovations often resulted from creative adaptation of existing technologies rather than completely new inventions – demonstrating how emergency conditions can accelerate practical innovation.

One significant innovation came in power management for emergency communications. Teams developed hybrid systems that combined conventional generators with vehicle alternator power, creating extended runtime for critical equipment when fuel supplies were severely limited. This approach, developed under the pressure of immediate need, has since evolved into more sophisticated hybrid power solutions now included in standard emergency kits.

This example illustrate how the urgency and constraints of disaster conditions can drive practical innovation that addresses immediate needs while evolving into more refined solutions for future response efforts.

The Future of Telecommunications Emergency Response

Looking ahead, several emerging trends and technologies will likely transform telecommunications emergency response in the coming years.

Integration of autonomous systems will increasingly complement human responders in disaster environments. This evolution includes:

- Autonomous drones that conduct initial infrastructure assessments without human control
- Self-deploying communications nodes that automatically position themselves for optimal coverage
- Robotic systems that perform hazardous installation tasks in unstable environments
- AI-driven operations centers that maintain service while human attention focuses on complex decisions

Rather than replacing human responders, these autonomous capabilities will extend what teams can accomplish, particularly during the critical early hours when resources are most limited.

Resilient network architectures designed specifically for disaster conditions represent another important direction. These approaches move beyond adaptation of commercial networks to purpose-built designs for emergency resilience, including:

- Self-healing mesh technologies that automatically reconfigure to maintain service despite infrastructure damage

- Hybrid terrestrial-aerial-satellite systems that seamlessly transition between available transmission paths
- Dynamic spectrum utilization that adapts to available frequencies in disrupted environments
- Distributed core functions that eliminate single points of failure in network architecture

These architectures promise faster, more reliable service restoration following disasters, particularly for critical emergency communications needs.

Predictive deployment based on advanced risk modeling may significantly improve response times in the future. These systems would:

- Analyze environmental data, infrastructure vulnerability, and threat patterns
- Predict likely damage patterns before disasters occur
- Pre-position resources based on these predictions
- Create customized response plans for specific anticipated scenarios

By positioning resources before disasters rather than reacting afterward, these approaches could dramatically reduce the time required to restore essential communications.

Edge computing for emergency resilience represents a promising approach for maintaining critical services even when backhaul connectivity is compromised. By distributing computing resources throughout the network edge, these architectures allow:

- Local services to continue functioning when separated from central systems

- Reduced bandwidth requirements for essential applications
- Faster service restoration for priority functions
- Graceful degradation rather than complete failure when network sections are isolated

This distributed approach aligns particularly well with the fragmented network environments common in disaster zones.

International standardization of emergency telecommunications continues to advance through organizations like the International Telecommunication Union (ITU) and the Emergency Telecommunications Cluster (ETC). These efforts focus on:

- Common technical standards for emergency communications equipment
- Harmonized regulatory frameworks for disaster telecommunications
- Standardized resource sharing protocols between countries and organizations
- Coordinated training and certification for emergency telecommunications personnel

As these standards mature, they promise to improve interoperability, reduce deployment barriers, and enhance the global community's collective capability to respond to major disasters.

The telecommunications emergency response capability of the future will likely combine these advancing technologies with the human expertise and adaptability that remain essential in disaster environments. Organizations that effectively integrate technological innovation with skilled personnel and proven response frameworks will be best positioned to provide critical communications when they are most urgently needed.

CHAPTER 8: AFTER THE STORM: RECOVERY AND LEARNING

The successful conclusion of emergency operations doesn't mark the end of the disaster response cycle but rather a transition to the crucial phases of recovery, evaluation, and preparation for future events. This chapter explores how to effectively manage this transition while capturing the invaluable learning that emerges from each emergency experience.

Transition to Recovery Operations

The shift from emergency response to recovery represents a critical juncture that requires careful management to maintain service continuity while transitioning to more sustainable long-term solutions.

Emergency to normal operations handoff must be carefully planned and executed to prevent service disruptions during the transition. This process typically includes:

- Detailed documentation of all emergency installations including equipment configurations, site information, and known issues
- Joint inspections involving both emergency response and regular operations personnel
- Phased transition that maintains redundancy until permanent systems are fully functional
- Clear transfer of responsibility through formal handover procedures

- Ongoing consultation availability from emergency teams during the early recovery period

Effective handoffs ensure that the progress made during emergency operations continues seamlessly into the recovery phase without creating new service gaps or technical problems.

Progressive stand-down procedures determine how emergency resources are demobilized as recovery progresses. Rather than abruptly terminating emergency operations, effective organizations implement a phased approach that includes:

- Criteria-based reductions in emergency staffing and equipment
- Scheduled verification that permanent solutions are functioning properly before removing temporary systems
- Retention of quick-response capability even as major operations wind down
- Clear communication about transition timelines to all stakeholders
- Recognition and decompression activities for departing emergency personnel

This measured approach prevents premature withdrawal of essential resources while allowing for efficient redeployment as the situation stabilizes.

Resource redeployment planning addresses how emergency assets will be handled as operations conclude. This planning includes:

- Inspection and maintenance requirements for all equipment before storage or redeployment
- Restocking of consumed supplies and replacement of damaged components

- Prioritization for reconstituting emergency caches in preparation for future events
- Transportation and logistics for moving equipment from the disaster zone
- Security procedures for equipment during the transition period

Proper resource management during this phase ensures that emergency response capability is quickly reconstituted rather than being depleted by the current operation.

Documentation and record-keeping take on particular importance during the transition to recovery. Comprehensive documentation should include:

- Detailed inventory of all installed emergency equipment with disposition plans
- Complete configuration information for all systems
- Maps and site information for all emergency installations
- Contact information for key stakeholders and decision-makers
- Chronological record of major activities and milestones

This documentation provides essential reference information for both recovery operations and future emergency planning.

Financial tracking and reporting must be finalized as emergency operations conclude. This process typically includes:

- Compilation of all costs associated with the emergency response
- Allocation of expenses to appropriate categories for reimbursement

- Submission of required documentation for insurance claims
- Preparation of regulatory filings related to emergency expenditures
- Development of financial summaries for organizational leadership

Proper financial management during this phase ensures appropriate cost recovery while providing valuable data for future emergency budgeting and planning.

After-Action Review Process

The after-action review (AAR) transforms emergency experience into organizational learning that improves future response capability. This process must be more than a ceremonial exercise – it should be a structured, honest assessment that drives meaningful improvement.

Structured debriefing methodology provides the foundation for effective after-action review. This methodology typically includes:

- Multiple sessions addressing different aspects of the response (technical, logistical, coordination, etc.)
- Participation from all key roles and levels within the response organization
- Neutral facilitation that encourages candid discussion
- Focus on systems and processes rather than individual blame
- Balanced examination of both successes and shortcomings

The methodology should create psychological safety that enables honest assessment while maintaining sufficient structure to ensure comprehensive coverage of all relevant aspects of the response.

Lessons identified vs. lessons learned represents a crucial distinction in the after-action process. Lessons identified are observations about what worked or didn't work during the response, while lessons learned are actual changes implemented as a result. Effective organizations track both, recognizing that value comes not from identification alone but from the implementation of improvements based on experience. This tracking typically includes:

- Systematic categorization of identified lessons

- Assessment of patterns and recurring themes
- Prioritization based on operational impact
- Assignment of responsibility for implementation
- Verification that changes have been effectively implemented

This disciplined approach ensures that insights gained during emergency operations translate into tangible improvements in capability.

Implementation planning for improvements transforms review insights into concrete action. Effective implementation planning includes:

- Specific, measurable objectives for each improvement
- Clear ownership and accountability for implementation
- Realistic timelines that recognize resource constraints
- Integration with existing improvement programs and processes
- Defined metrics for evaluating implementation success

This planning prevents after-action insights from becoming merely interesting observations that fail to drive actual change.

Recognition and appreciation processes acknowledge the extraordinary efforts of emergency response personnel while reinforcing organizational values. These processes might include:

- Formal recognition events that highlight key contributions
- Letters of appreciation from senior leadership

- Special recognition for exemplary performance or innovation
- Team celebration activities that build camaraderie
- Appropriate sharing of success stories with broader stakeholders

While primarily focused on appropriate acknowledgment of personnel contributions, these processes also reinforce the behaviors and approaches the organization wishes to encourage in future responses.

Interagency feedback collection extends the learning process beyond organizational boundaries to include external partners and stakeholders. This feedback provides valuable perspective on how the organization's response integrated with broader emergency efforts and might include:

- Structured feedback sessions with partner agencies
- Surveys of key stakeholders regarding response effectiveness
- Participation in multi-agency after-action reviews
- Integration of external perspectives into internal improvement planning
- Sharing of relevant learning with the broader emergency response community

This expanded perspective helps identify improvement opportunities that might not be apparent from an internal perspective alone.

Long-term Planning Implications

Beyond immediate improvements to emergency response capability, disaster experiences often reveal opportunities for more fundamental enhancements to infrastructure, processes, and organizational approaches.

Infrastructure hardening priorities emerge from the patterns of failure observed during disasters. The after-action review should systematically identify these patterns and develop recommendations for long-term infrastructure improvements, which might include:

- Strengthening of vulnerable tower structures
- Relocation of critical facilities from high-risk areas
- Implementation of redundant power systems at key sites
- Enhancement of backhaul diversity to reduce single points of failure
- Deployment of more resilient equipment in vulnerable locations

While immediate post-disaster budgets may not support all these improvements, documenting the priorities provides essential input for long-term capital planning.

Pre-positioning strategies revision often results from observing which emergency resources proved most valuable and where deployment delays created significant impacts. Enhancements to pre-positioning might include:

- Redistribution of emergency caches based on observed access challenges

- Expansion of certain resource types that proved especially valuable
- Development of new cache locations to address identified gaps
- Enhancement of cache management systems to ensure readiness
- Implementation of improved inventory tracking for emergency resources

These refinements increase both the effectiveness and efficiency of future emergency deployments.

Training program adjustments should directly reflect the skills and capabilities that proved most valuable or most lacking during the actual emergency. Potential enhancements might include:

- Development of new training modules addressing identified skill gaps
- Increased emphasis on areas where performance showed room for improvement
- Creation of specialized training for roles that faced particular challenges
- Enhancement of simulation exercises based on actual disaster conditions
- Implementation of joint training with partner organizations where coordination was challenging

These adjustments ensure that training programs evolve based on real-world experience rather than theoretical assumptions.

Equipment and resource reconsideration examines the tools and materials used during the emergency response, identifying both highly effective items that should be expanded and those that proved less valuable than expected. This review might lead to:

- Acquisition of new equipment types that would have improved response effectiveness
- Reduction or elimination of resources that saw little use relative to their logistics burden
- Modifications to standard kit compositions based on actual usage patterns
- Development of improved specifications based on field performance
- Exploration of emerging technologies that could address identified limitations

This continuous refinement of the resource inventory ensures that future responses benefit from both current experience and technological advancement.

Procedure and protocol updates translate operational lessons into formal guidance for future responses. These updates might address:

- Activation decision criteria and processes
- Assessment methodologies and reporting formats
- Prioritization frameworks for resource allocation
- Coordination mechanisms with partner organizations
- Communication procedures for various stakeholder groups

By systematically incorporating lessons from each emergency into formal procedures, organizations create an continuously improving foundation for future responses.

Case Study: Post-Hurricane Improvements

The response to major hurricanes has driven significant evolution in telecommunications emergency capabilities, with each event revealing both strengths to maintain and opportunities for enhancement.

Field Observations: Learning from Hurricane Response

Over my years monitoring telecommunications emergency response, I've observed how learning from hurricane response operations has driven substantial improvements in both technology and process. These enhancements provide valuable examples of how organizations can turn disaster experience into increased capability.

One of the most significant improvements came in power management systems. Early hurricane responses often struggled with power sustainability for emergency communications equipment, particularly when commercial power restoration took longer than anticipated. Learning from these experiences, telecommunications providers developed much more sophisticated approaches including:

- Hybrid power systems combining generators, solar, and advanced battery management
- Fuel consumption monitoring that provided early warning of potential shortages
- Power prioritization systems that automatically shed non-essential loads to maintain core functions

- Standardized interfaces that allowed for field-swapping of different power sources as availability changed

These enhancements directly addressed one of the most common failure points in previous response efforts, significantly increasing the sustainability of emergency communications in subsequent disasters.

Coordination mechanisms between telecommunications providers and government agencies also evolved substantially based on hurricane response experiences. Where early responses often featured fragmented efforts and communication challenges, later responses benefited from:

- Pre-established liaison positions embedded with emergency management agencies
- Standardized reporting formats that aligned with government emergency systems
- Joint planning processes that integrated telecommunications restoration with broader recovery efforts
- Clear escalation paths for resolving resource conflicts or priority disagreements

These improved coordination mechanisms enhanced both the effectiveness and efficiency of telecommunications restoration, ensuring that limited resources achieved maximum impact within the overall emergency response.

Perhaps most importantly, I've observed significant evolution in how organizations approach team preparation and support based on hurricane response experiences. Improvements in this area include:

- More realistic pre-deployment briefings that better prepare teams for field conditions

- Enhanced psychological support resources available during and after deployment
- More structured rotation planning to prevent fatigue during extended operations
- Improved family support programs that reduce stress for deployed personnel

These human-focused enhancements recognize that even the best technology and processes depend ultimately on the effectiveness of the people implementing them – particularly under the challenging conditions of disaster response.

Field-Proven Protocols: Company-Level Emergency Planning

Effective disaster response requires organizational preparation:

Emergency Committee Structure:

- Establish a disaster response committee including:
    - General Management
    - Engineering/Systems
    - Finance/Accounting
    - Human Resources
    - Material Resources
    - Sales (this department typically has the most human and material resources available in or near affected areas)

- Designate an internal communication channel open 24/7

Resource Planning:

- Maintain sufficient vehicles for emergency response, fully fueled and in good mechanical condition
- Review insurance policies
- Recognize that these vehicles will be unavailable for normal operations in the short and medium term
- Review available personnel for the coming weeks, consider rotation for disaster duty while maintaining minimum staff for regular operations
- Assess material inventory, define central supply site and complementary warehouses
- Maintain sufficient equipment, materials, and tools for at least two months of contingency operations

Government Coordination:

- The General Director should establish contact with telecommunications regulatory authorities
- Open a communication channel with these agencies, informing them of installed/available capacity both at the affected site and in neighboring populations or states
- Be prepared to deploy personnel as soon as authorities request it and confirm that security conditions exist for working in the area
- Place official documentation in all vehicles, showing that personnel are dedicated to restoring telecommunications in the affected area

Media Relations:

- When receiving requests from media, journalists, or interview requests about the situation and actions taken, refer only to the company's actions
- Do not provide opinions, estimates, or rumors about the situation
- Always refer to official telecommunications authorities for real-time information

These field-tested protocols, developed through actual disaster response operations, provide practical guidance that can help telecommunications teams operate effectively under the most challenging conditions.

CHAPTER 9: BUILDING A SUSTAINABLE EMERGENCY RESPONSE CAPABILITY

Beyond the immediate requirements of disaster response, telecommunications organizations must develop sustainable emergency capabilities that can be maintained over time despite competing priorities and resource constraints. This chapter explores the organizational foundations that support effective emergency response as an ongoing capability rather than a periodic scramble.

Making the Business Case

Telecommunications executives and boards naturally focus on mainstream business objectives like market growth, customer satisfaction, and financial performance. Building support for significant investment in emergency response capabilities requires a compelling business case that connects these capabilities to core organizational priorities.

Risk management represents perhaps the most straightforward justification for emergency response investment. Telecommunications providers face quantifiable financial risks from service disruptions, including:

- Revenue loss during outages
- Customer compensation requirements
- Regulatory penalties for failure to restore essential services
- Permanent customer loss due to prolonged disruptions

- Reputational damage that affects future business opportunities

A structured risk analysis can translate these potential losses into financial terms, comparing them against the required investment in emergency response capabilities to demonstrate a favorable return on investment even in purely financial terms.

Regulatory compliance frequently mandates specific emergency response capabilities, particularly for providers of essential services. These requirements typically include:

- Minimum restoration timeframes for various service categories
- Required emergency equipment and capabilities
- Participation in industry mutual aid agreements
- Regular testing and reporting on emergency preparedness
- Coordination with government emergency management agencies

While compliance-driven investments might be considered minimum requirements rather than strategic capabilities, they provide a foundation that can be enhanced to deliver additional organizational benefits.

Corporate social responsibility aspects of emergency response are increasingly important to both customers and investors. Organizations that demonstrate effective commitment to community support during disasters often experience:

- Enhanced public perception and brand reputation
- Improved relationships with government agencies and regulators

- Higher employee engagement and retention
- Customer preference based on perceived reliability and community

Competitive advantage considerations often provide the most compelling justification for investment beyond minimum requirements. Organizations with superior emergency response capabilities can:

- Maintain service when competitors experience outages, creating opportunities for customer acquisition
- Restore service more quickly after widespread disasters, enhancing customer loyalty
- Support critical enterprise and government customers during emergencies, strengthening high-value relationships
- Demonstrate operational excellence that differentiates the organization in competitive procurement processes

These competitive advantages translate into tangible business benefits through customer acquisition, retention, and relationship development that extend well beyond the emergency period itself.

Brand and reputation impact may represent the most significant long-term business justification for emergency response investment. Telecommunications organizations that perform effectively during disasters often experience lasting reputational benefits that influence customer decisions for years afterward. Conversely, those that fail to meet community expectations during crises may face persistent brand damage that affects market position long after service is restored.

Customer retention during disasters provides immediate financial justification for emergency response capabilities. Research consistently shows that service resilience during disasters significantly influences customer loyalty, with substantial differences in churn rates between providers based on their performance during emergencies. The lifetime value of customers retained through effective disaster response typically far exceeds the investment required for emergency capabilities.

Resource Allocation Models

Developing sustainable emergency response capability requires thoughtful allocation of limited organizational resources. Several models exist for structuring this capability, each with distinct advantages and limitations.

Dedicated vs. distributed team structures represent a fundamental structural decision. Dedicated emergency teams maintain specialized personnel whose primary responsibility is emergency preparedness and response. These teams typically demonstrate superior technical capabilities and response readiness but require significant ongoing investment that doesn't directly contribute to normal operations. Distributed models embed emergency responsibilities within regular operational teams, activating these capabilities only during emergencies. This approach utilizes resources more efficiently during normal operations but may result in diminished emergency capabilities as daily priorities consistently overshadow emergency preparation.

Most effective organizations implement hybrid models that combine elements of both approaches. These typically feature a small core of dedicated emergency specialists who maintain equipment, develop procedures, conduct training, and lead actual emergency operations. This core team is augmented during emergencies by operational personnel who receive specialized training and participate regularly in emergency exercises. This model balances readiness with resource efficiency while still maintaining critical specialized capabilities.

Equipment investment strategies involve similar considerations. Some organizations maintain extensive dedicated emergency equipment that remains unused during normal operations, ensuring maximum availability during emergencies but creating significant idle resource costs. Others rely primarily on repurposing operational equipment during emergencies, minimizing dedicated emergency resources but potentially compromising both normal operations and emergency response when disasters occur.

Effective approaches typically involve strategic redundancy—maintaining critical emergency-specific equipment while designing operational systems with enough flexibility and redundancy to support emergency repurposing when needed. This balanced approach ensures essential emergency capabilities while maximizing resource utilization during normal operations.

Training and readiness budgeting represents another critical resource allocation decision. Organizations must determine how much time operational personnel will dedicate to emergency training and exercises rather than their primary responsibilities. This allocation directly affects both emergency readiness and normal operational capacity.

The most sustainable approaches integrate emergency skills development into broader professional development programs, ensuring that emergency capabilities enhance rather than compete with core operational skills. For example, training in improvised repairs and degraded mode operations often improves troubleshooting capabilities valuable during normal operations, while emergency leadership development contributes to broader management capabilities.

Mutual aid agreements and cost-sharing arrangements with other organizations can significantly enhance emergency capabilities while reducing individual organizational investment. These agreements include:

- Shared emergency equipment caches maintained jointly by multiple providers
- Coordinated training programs that distribute development costs across organizations
- Personnel sharing agreements that allow specialized resources to support multiple providers
- Joint procurement arrangements that reduce equipment costs through volume purchasing
- Regional coordination centers that distribute overhead costs across multiple participants

These collaborative approaches recognize that while providers may compete in normal operations, they share common interests in effective disaster response and can achieve better outcomes through cooperation than through individual efforts.

Insurance and risk transfer mechanisms complement direct emergency response capabilities by addressing financial impacts that cannot be prevented through operational measures. Comprehensive risk management strategies typically include:

- Specialized telecommunications service interruption insurance
- Business continuity insurance covering extraordinary expenses during emergencies

- Equipment damage coverage with expedited replacement provisions
- Liability protection for emergency operations in non-standard conditions
- Parametric insurance that provides immediate funding based on disaster characteristics

These financial mechanisms ensure that organizations maintain the resources needed for effective emergency response even when facing substantial disaster-related costs.

Maintaining Readiness

Perhaps the greatest challenge in telecommunications emergency response isn't developing initial capabilities but maintaining them over time amid competing priorities and the perception that major disasters are rare events. Sustained readiness requires systematic approaches that keep emergency capabilities viable without constant crisis conditions.

Regular training and exercise schedules provide the foundation for maintained readiness. These activities should follow a structured calendar that includes:

- Monthly skill maintenance activities for core emergency team members
- Quarterly technical exercises testing specific emergency equipment and procedures
- Semi-annual functional exercises focused on coordination and decision-making
- Annual full-scale simulations that test all aspects of emergency response
- Bi-annual participation in industry or government-led regional exercises

This graduated approach maintains skills and coordination capabilities while distributing the resource impact throughout the year rather than concentrating it in major exercises.

Equipment testing and rotation prevents the deterioration of emergency resources during storage. Effective programs typically include:

- Monthly operational testing of all emergency power systems

- Quarterly deployment of satellite and portable communications systems
- Semi-annual full technical testing of all emergency equipment
- Equipment rotation that periodically introduces emergency resources into operational use
- Preventive maintenance schedules specific to each equipment type

These activities ensure that emergency equipment will function when needed while identifying and addressing any deterioration before it affects emergency capabilities.

Team member qualification maintenance addresses the natural degradation of skills and knowledge over time. Comprehensive qualification programs typically include:

- Individual certification requirements for specific emergency skills
- Regular assessment of both technical and operational capabilities
- Recertification requirements based on demonstrated competence
- Tracking systems that monitor qualification status across the organization
- Remediation programs that address identified qualification gaps

This systematic approach prevents the gradual erosion of emergency response capabilities that otherwise naturally occurs between major activations.

Knowledge management systems preserve critical information and insights that might otherwise be lost through personnel changes or memory degradation. Effective systems include:

- Detailed documentation of all emergency procedures and protocols
- Lessons learned databases capturing insights from both exercises and actual emergencies
- Technical libraries containing emergency equipment information and field-expedient solutions
- Video libraries demonstrating key emergency techniques and procedures
- Explicit documentation of tacit knowledge from experienced emergency personnel

These systems ensure that organizational emergency capabilities don't depend solely on the knowledge of specific individuals who may not always be available when disasters occur.

Early warning and monitoring capabilities help organizations transition from normal operations to emergency response before infrastructure damage occurs. Effective systems typically include:

- Integration with government weather and geological monitoring services
- Participation in emergency management information sharing networks
- Automated alerting systems for both natural and human-caused threats
- Escalation procedures that increase readiness as threat indicators intensify

- Predictive analytics that identify potential emergency conditions from system performance data

These capabilities allow for proactive preparation and potentially pre-positioning of resources before disasters strike, significantly enhancing response effectiveness.

Vendor and partner relationship management ensures that external resources will be available during emergencies. Sustainable programs include:

- Emergency provisions in standard vendor contracts
- Priority service agreements with critical suppliers
- Regular verification of vendor emergency capabilities
- Joint exercises with key equipment and service providers
- Established emergency procurement procedures and authorizations

These arrangements prevent dependency on vendors who cannot deliver during emergency conditions while ensuring rapid access to additional resources when needed.

Integration with Business Continuity

Telecommunications emergency response represents one aspect of broader organizational resilience that includes business continuity, disaster recovery, and crisis management. These capabilities are most effective when integrated rather than developed and maintained as separate programs.

Aligning emergency response with continuity plans ensures that technical restoration efforts support critical business functions. This alignment includes:

- Prioritization frameworks that consider both technical and business impacts
- Recovery time objectives that reflect business requirements rather than purely technical considerations
- Resource allocation approaches that balance infrastructure restoration with business process recovery
- Integrated activation procedures that coordinate technical and business recovery efforts
- Joint governance structures that manage both dimensions of organizational recovery

This integration ensures that limited resources are directed toward the most business-critical services rather than following purely technical priorities that might not align with organizational needs.

Complementary capabilities development maximizes the value of investments across resilience programs. Effective approaches include:

- Cross-training that enables personnel to support multiple aspects of organizational recovery

- Multi-purpose equipment that supports both technical and business recovery needs
- Joint documentation systems that maintain information for all resilience programs
- Shared monitoring and alerting systems that support both technical and business continuity
- Integrated vendor management that addresses all aspects of emergency support

This coordinated approach increases overall organizational resilience while reducing the total investment required compared to developing separate, independent capabilities.

Shared resources and efficiencies emerge from integrated resilience programs. These typically include:

- Combined emergency operations centers that support all aspects of organizational response
- Unified communications systems for all emergency and recovery activities
- Joint training and exercise programs that address multiple resilience dimensions
- Consolidated vendor relationships that leverage larger volumes for better terms
- Integrated technology platforms that support all resilience functions

These shared approaches reduce duplication while creating a more coordinated organizational response to emergencies.

Integrated testing and exercises provide more realistic preparation while reducing the total time and resources required. Effective programs typically include:

- Scenarios that test both technical restoration and business recovery
- Role players representing all organizational functions and external stakeholders
- Performance metrics that address both technical and business objectives
- Debriefing processes that capture learning across all resilience dimensions
- Improvement planning that considers the full spectrum of emergency capabilities

These integrated activities ensure that all aspects of organizational resilience work together effectively while identifying coordination issues that might not emerge in function-specific exercises.

Unified governance models provide oversight and direction for all resilience programs. Effective governance typically includes:

- Executive leadership with responsibility for all resilience functions
- Integrated planning processes that address all emergency and recovery needs
- Consolidated budgeting that ensures appropriate resource allocation across programs
- Coordinated policy development that creates consistent direction for all resilience activities

- Joint performance reporting that provides comprehensive visibility into organizational preparedness

This governance approach prevents the fragmentation of resilience efforts while ensuring appropriate executive attention to emergency preparedness despite competing priorities.

Regulatory Considerations

Telecommunications emergency response operates within a complex regulatory environment that both mandates certain capabilities and potentially constrains emergency actions. Understanding and effectively navigating these regulatory considerations is essential for sustainable emergency response.

Compliance frameworks establish minimum requirements for telecommunications emergency preparedness and response. These typically include:

- Required restoration timeframes for different service categories
- Mandatory reporting during service disruptions
- Participation in industry coordination mechanisms
- Minimum backup power capabilities for critical infrastructure
- Regular testing and certification of emergency capabilities

While compliance represents a baseline rather than a comprehensive approach to emergency response, it provides an essential foundation that ensures basic capabilities across the industry.

Government partnership opportunities often complement regulatory requirements with voluntary collaborative programs. These may include:

- Joint planning activities with emergency management agencies
- Participation in government-sponsored exercises and training

- Information sharing about threats and vulnerabilities
- Technical assistance programs for emergency capability development
- Grant funding for specific emergency response enhancements

These partnerships can significantly enhance organizational capabilities while building relationships that prove valuable during actual emergencies.

Industry cooperation frameworks provide mechanisms for coordinated response that maximizes collective impact. These typically include:

- Formal mutual aid agreements for resource sharing during emergencies
- Industry emergency coordination centers that facilitate collaborative response
- Standardized emergency procedures and terminology across providers
- Joint training and certification programs that ensure interoperability
- Shared situation assessment and information distribution mechanisms

These cooperative approaches recognize that disasters affect entire regions rather than individual providers and that the most effective response requires coordinated industry action.

International response coordination becomes increasingly important as telecommunications infrastructure spans national boundaries and disasters affect multiple countries simultaneously. Key considerations include:

- Cross-border mutual aid agreements for emergency response
- Harmonized regulatory frameworks that facilitate international assistance
- Customs and immigration provisions for emergency personnel and equipment
- Standardized equipment specifications that ensure interoperability
- Multinational exercises that practice international coordination

These international mechanisms are particularly important for organizations operating across multiple countries but benefit all providers during widespread disasters that exceed national response capabilities.

Standardization opportunities emerge from both regulatory requirements and industry initiatives. Effective standardization typically addresses:

- Emergency equipment specifications that ensure interoperability
- Communication protocols for emergency coordination
- Training and certification requirements for emergency personnel
- Documentation formats for emergency plans and procedures
- Performance metrics for emergency response effectiveness

These standards reduce both the cost and complexity of emergency response while facilitating coordination across organizational boundaries during disasters.

MARCOS DUARTE RIVERO

CONCLUSION: COMMUNICATIONS AS A LIFELINE

Throughout this guide, we've explored the complex, challenging, and essential work of telecommunications emergency response. From team structure and equipment selection to psychological support and learning processes, we've examined the components that enable effective restoration of communications when disaster strikes.

The Vital Role of Telecommunications in Disaster Resilience

As we've seen throughout this book, telecommunications infrastructure provides far more than convenience during disasters – it serves as a critical lifeline that enables all other aspects of emergency response and community recovery.

When disaster strikes, communications systems enable emergency coordination that saves lives. First responders depend on reliable communications to direct resources where they're most needed, coordinate complex operations across multiple agencies, and maintain situational awareness in rapidly changing conditions. Without these capabilities, even the most well-resourced emergency services struggle to operate effectively.

Beyond emergency services, telecommunications connects affected communities with vital information about safety measures, resource availability, and recovery progress. This information reduces uncertainty and anxiety while enabling individuals to make informed decisions about their own situation.

During prolonged recovery periods, communications systems also support essential services like remote healthcare, emergency financial assistance, and educational continuity.

For individuals facing disaster, telecommunications provides the precious ability to connect with family and support networks. The psychological value of these connections cannot be overstated – they provide reassurance of safety, emotional support during traumatic experiences, and practical assistance for recovery needs. As we observed during Hurricane Otis, the community connectivity points established in public squares became gathering places where the human dimension of disaster response became most visible.

In our increasingly connected society, telecommunications has evolved from a convenience to an essential service that communities depend upon for both everyday functioning and disaster resilience. The telecommunications responders who restore these services during crisis don't merely reestablish technology – they reconnect the human networks that sustain communities through their most challenging moments.

How Prepared Response Teams Save Lives and Communities

As this guide has demonstrated, the effectiveness of telecommunications restoration during disasters depends fundamentally on preparation that occurs long before disaster strikes. Well-prepared response teams save lives and communities through several critical capabilities.

Speed of response represents perhaps the most crucial advantage of prepared teams. Organizations with pre-positioned equipment, trained personnel, and established protocols can deploy communications solutions within hours rather than days – a difference that directly affects how many people receive timely emergency assistance. During Hurricane Otis, our ability to establish the first satellite connection at the Red Cross command post within hours of arrival dramatically improved coordination of the entire disaster response effort.

Sustainability of operations distinguishes truly effective emergency response. Teams with comprehensive logistical preparation, self-sufficiency training, and resource management systems can maintain communications services throughout extended disruptions rather than providing only temporary solutions. This sustained capability supports both immediate life-saving operations and the longer recovery process that communities require to regain stability.

Appropriate prioritization ensures that limited telecommunications resources achieve maximum humanitarian impact. Teams with established frameworks for determining where to deploy first can make these difficult decisions quickly and consistently, directing capabilities where they will save the most lives rather than following arbitrary or politically influenced allocation. During Hurricane Otis, the focus on establishing connectivity for emergency services coordination before addressing general public communications maximized the overall effectiveness of the response.

Coordination with broader emergency efforts multiplies the impact of telecommunications restoration. Teams trained to integrate with government agencies, other providers, and humanitarian organizations ensure that their technical capabilities support a coherent overall response rather than operating in isolation. This integrated approach creates synergies that benefit affected communities far beyond what telecommunications alone could provide.

Psychological support for affected populations often emerges as an unexpected but crucial contribution of telecommunications teams. By reconnecting isolated communities, enabling communication with family members, and restoring a sense of connection to the outside world, these teams address not just practical needs but also the profound psychological impact of disaster isolation. The visible relief on people's faces when they can finally call loved ones demonstrates this dimension powerfully.

The Ongoing Commitment to Emergency Readiness

The work of telecommunications emergency response doesn't end when a specific disaster recovery concludes – it continues through an ongoing commitment to readiness that ensures capability for future events. This commitment takes several forms that organizations must maintain even when no immediate threat is visible.

Regular training and simulation ensure that response teams maintain both technical skills and operational readiness between disasters. These activities should include realistic exercises that test the full range of capabilities under conditions approximating actual emergencies. Organizations that treat emergency training as a continuous process rather than an occasional event develop teams that perform effectively from the moment they deploy.

Equipment maintenance and technology evolution require constant attention to ensure that emergency resources incorporate both lessons from past deployments and advances in available solutions. This includes not just keeping existing equipment in deployment-ready condition but also evaluating and integrating new technologies that could enhance response capabilities. The telecommunications landscape evolves rapidly, and emergency response capabilities must evolve alongside it.

Relationship development with government agencies, other providers, and humanitarian organizations builds the foundation for effective coordination during actual emergencies. These relationships should be actively maintained through regular contact, joint planning activities, and shared training opportunities. The middle of a disaster is never the right time to exchange contact information or establish communication protocols – these must be developed during normal operations.

Organizational commitment at leadership levels sustains emergency response capability through changing business conditions and competing priorities. This commitment includes appropriate budget allocation, personnel assignments, and strategic emphasis even when emergency response may not seem immediately relevant to business objectives. Organizations that maintain this commitment distinguish themselves during disasters, demonstrating both technical capability and humanitarian leadership when communities need them most.

Individual preparation by team members embodies perhaps the most fundamental aspect of readiness. The telecommunications professionals who maintain their personal equipment, update their family emergency plans, and continuously develop their skills demonstrate a commitment that transcends organizational requirements. These individuals form the heart of effective emergency response – technical experts who understand that their skills serve a profound humanitarian purpose when disaster strikes.

Final Reflections from Hurricane Otis

The Hurricane Otis experience that has informed many examples throughout this guide offers several final reflections worth carrying forward.

First, the boundary between "normal" telecommunications work and emergency response is increasingly blurred in our connected society. The technicians who establish connectivity in underserved communities during regular operations develop skills and mindsets directly applicable to disaster response. Organizations that recognize this connection can build emergency capability while serving broader social purposes.

Second, the most effective emergency response combines technical expertise with genuine humanitarian commitment. The Hughes installation teams who worked tirelessly to establish those initial satellite connections for the Mexican Red Cross weren't merely performing technical tasks – they understood that their work directly enabled life-saving operations throughout the affected region. This understanding provided motivation and purpose that transcended normal professional responsibilities.

Third, telecommunications emergency response represents a unique form of international cooperation that transcends political boundaries. The Hurricane Otis response involved collaboration across national borders, between commercial competitors, and across disparate organizations with different missions – all united by the shared purpose of reconnecting affected communities. This collaborative spirit offers a model for addressing other challenges that similarly transcend traditional boundaries.

Finally, the Hurricane Otis experience reinforced that telecommunications emergency response is fundamentally about people rather than technology. While the technical details matter greatly for operational effectiveness, the ultimate purpose and measure of success lies in human impact – communities reconnected, families reassured of each other's safety, and emergency services coordinated to save lives.

A Call to Action for the Telecommunications Industry

As we conclude this guide, I offer a call to action for telecommunications organizations of all types and sizes. The capability to respond effectively when disaster strikes represents both an operational necessity and a profound social responsibility for our industry. Building this capability requires commitment, investment, and preparation that must begin long before the next disaster makes headlines.

First, recognize that emergency response capability isn't merely a regulatory requirement or corporate social responsibility initiative – it's an essential component of telecommunications service in a world where communities depend on connectivity for both everyday functioning and disaster resilience. Organizations that integrate this understanding into their core identity develop more effective responses while fulfilling their broader purpose in society.

Second, invest in the human dimension of emergency response alongside technical capabilities. This includes selecting personnel with appropriate psychological profiles, providing specialized training for disaster conditions, and developing support systems that sustain teams through the challenges of emergency deployment. Organizations that recognize emergency response as both a technical and human challenge build capabilities that perform effectively under the most difficult conditions.

Third, participate actively in the broader emergency response community through government partnerships, industry collaboration, and information sharing. Telecommunications emergency response never happens in isolation – it functions within an ecosystem of organizations addressing different aspects of disaster impact. By engaging with this community during normal operations, telecommunications providers enhance both their own effectiveness and the overall disaster response system.

Fourth, commit to continuous learning and improvement based on each emergency experience. Every disaster provides invaluable lessons about what works, what doesn't, and what could work better with appropriate adjustment. Organizations that systematically capture these lessons and incorporate them into future preparations build capabilities that continuously evolve rather than merely repeating past approaches.

Finally, remember that telecommunications emergency response represents some of the most meaningful work our industry performs. When communities face their most challenging moments, the telecommunications professionals who restore connectivity don't merely reestablish technical services – they reconnect the human networks that sustain communities through crisis and recovery. This work embodies the highest purpose of our technology and our industry.

As we face a future likely to include more frequent and severe disasters due to climate change, infrastructure interdependence, and other factors, the telecommunications industry's emergency response capability takes on even greater importance. By developing this capability through the approaches outlined in this guide, telecommunications organizations prepare not just for their own business continuity but for their essential role in community resilience during the most challenging circumstances.

The time to build this capability isn't when disaster makes headlines, but now – through systematic preparation, continuous improvement, and enduring commitment to connecting communities when they need it most.

CHECKLISTS

MARCOS DUARTE RIVERO

TELECOMMUNICATIONS INSTALLER/TECHNICIAN

EMERGENCY PREPAREDNESS CHECKLIST

PERSONAL READINESS

- ☐ Maintain a 24-hour go-bag in your vehicle at all times
- ☐ Keep a 72-hour deployment kit at home/office ready for immediate use
- ☐ Carry 5-day supply of personal medications on your person during deployments
- ☐ Have current medical clearance and vaccinations documented
- ☐ Establish and regularly update family emergency plan for your absence
- ☐ Ensure family members understand deployment protocols and communication expectations
- ☐ Set up automatic bill payments and financial arrangements for deployment periods
- ☐ Maintain backup personal communication device (satellite phone/communicator)
- ☐ Complete psychological preparedness training for disaster conditions
- ☐ Know and practice stress management techniques for high-pressure situations

TECHNICAL PREPAREDNESS

- ☐ Complete all required emergency response training certifications
- ☐ Maintain proficiency in manual configuration of all emergency equipment
- ☐ Practice equipment deployment under time pressure regularly
- ☐ Develop expertise in at least one alternative power solution
- ☐ Maintain personal tool kit for emergency deployments
- ☐ Learn field-expedient repair techniques for common equipment
- ☐ Practice operating in degraded mode without advanced features
- ☐ Maintain physical fitness required for disaster deployment
- ☐ Complete first aid and safety training for disaster environments
- ☐ Participate regularly in emergency simulation exercises

DEPLOYMENT ESSENTIALS

- ☐ Three changes of weather-appropriate clothing
- ☐ 10 liters of potable water and 4 liters of electrolyte drinks
- ☐ 5-day supply of non-perishable food (minimum 5 energy bars per day)
- ☐ Personal protective equipment (gloves, mask, helmet, eye protection)
- ☐ Sunscreen and insect repellent
- ☐ Waterproof emergency medical kit with:
    - ☐ Valid ID and emergency contact information
    - ☐ Basic first aid supplies
    - ☐ Water purification tablets
    - ☐ Flashlight with extra batteries
    - ☐ Multi-tool or basic tools
    - ☐ Emergency blanket
    - ☐ Dust mask and work gloves
- ☐ Sleeping bag or blanket
- ☐ Personal hygiene kit
- ☐ Portable phone charger and extra batteries
- ☐ Maps of deployment area (physical copies)
- ☐ Documentation of emergency protocols and contacts

SAFETY PRINCIPLES

- ☐ Commit to the safety hierarchy: "First me, then me, then me, then my teammate, then my vehicle, then my equipment, finally the installation site"
- ☐ Never work in excessive wind, rain, or electrical storms
- ☐ Never travel to areas without confirmed minimum safety conditions
- ☐ Never drive at night in affected areas
- ☐ Never work in isolation; always maintain buddy system
- ☐ Always notify base when departing or returning from field locations
- ☐ Verify vehicle condition before each deployment
- ☐ Plan for extended deployment with no services available
- ☐ Establish clear arrival and departure routes for each location
- ☐ Use maps and compass for navigation when digital systems are unavailable

INSTALLATION PRIORITIES

- ☐ Understand the established priority order for connectivity restoration:
    1. Emergency service headquarters (Red Cross, Fire, Police, Medical, National Guard)
    2. Hospitals and medical facilities
    3. Emergency operations centers and command posts
    4. Shelters and humanitarian aid centers
    5. Community connectivity hubs in public areas
- ☐ Document all installations with complete configuration details
- ☐ Establish communication schedule with base for each installation
- ☐ Verify power sustainability for each installation
- ☐ Conduct security assessment before each installation
- ☐ Provide basic operation training to on-site personnel
- ☐ Set clear expectations about capabilities and limitations
- ☐ Document all improvised solutions for knowledge sharing

Remember: Your technical expertise reconnects communities and enables all other aspects of disaster response. Your preparation directly impacts how quickly vital communications can be restored when people need them most.

MARCOS DUARTE RIVERO

GOVERNMENT / TELECOMMUNICATIONS AGENCY

EMERGENCY PREPAREDNESS CHECKLIST

REGULATORY FRAMEWORK AND AUTHORITY

- ☐ Establish clear emergency telecommunications regulations and statutory authority
- ☐ Define emergency telecommunications service provider obligations during disasters
- ☐ Implement priority telecommunications service regulations for critical users
- ☐ Create special authorization procedures for emergency equipment deployment
- ☐ Develop frequency allocation procedures for emergency operations
- ☐ Establish regulatory waivers for disaster conditions with clear activation criteria
- ☐ Define cross-border emergency telecommunications assistance framework
- ☐ Implement critical infrastructure designation for key telecommunications assets
- ☐ Create emergency spectrum management protocols

- □ Develop mandatory service restoration timeline requirements
- □ Establish mandatory reporting requirements during outages
- □ Define penalties for non-compliance with emergency regulations

COORDINATION MECHANISMS

- ☐ Establish National Emergency Telecommunications Coordination Center
- ☐ Create formal liaison positions with emergency management agencies
- ☐ Develop coordination protocols with military communications units
- ☐ Implement regular industry coordination meetings on emergency preparedness
- ☐ Create database of emergency contacts for all telecommunications providers
- ☐ Establish secure communication channels with critical infrastructure operators
- ☐ Develop international coordination mechanisms for cross-border disasters
- ☐ Create formal relationships with amateur radio and volunteer communications groups
- ☐ Establish coordination protocols with electricity/power providers
- ☐ Implement joint planning processes with telecommunications providers
- ☐ Develop information sharing framework for threat and vulnerability data
- ☐ Create standardized emergency communications protocols across agencies

RESOURCE MANAGEMENT

- ☐ Maintain national inventory of emergency telecommunications resources
- ☐ Establish strategic telecommunications equipment reserves
- ☐ Create resource allocation framework for national-level disasters
- ☐ Develop emergency procurement procedures for telecommunications equipment
- ☐ Establish equipment sharing and mutual aid protocols among providers
- ☐ Implement satellite communications capacity reservation for emergencies
- ☐ Create transportation priority system for telecommunications equipment
- ☐ Develop fuel allocation protocols for telecommunications power systems
- ☐ Establish emergency funding mechanisms for rapid telecommunications restoration
- ☐ Create import facilitation procedures for international emergency equipment
- ☐ Develop security resource allocation for telecommunications infrastructure protection
- ☐ Implement emergency telecommunications workforce credentialing system
- ☐

OPERATIONAL PREPAREDNESS

- ☐ Create national telecommunications vulnerability assessment program
- ☐ Develop service prioritization framework for restoration activities
- ☐ Establish standardized damage assessment methodology across providers
- ☐ Implement critical facilities telecommunications redundancy requirements
- ☐ Create telecommunications service continuity standards for essential services
- ☐ Develop backup systems testing and verification program
- ☐ Establish minimum power backup requirements for critical infrastructure
- ☐ Create emergency telecommunications deployment authorization system
- ☐ Develop technical standards for emergency communications equipment
- ☐ Implement national emergency alert system testing and verification
- ☐ Create telecommunications service impact reporting system
- ☐ Develop field deployment coordination mechanism for multiple providers

EXERCISE AND TRAINING PROGRAM

- ☐ Establish national telecommunications emergency exercise program
- ☐ Conduct quarterly coordination exercises with telecommunications providers
- ☐ Implement annual full-scale telecommunications failure simulation
- ☐ Create joint training programs with international counterparts
- ☐ Develop standardized telecommunications emergency response certifications
- ☐ Establish regional exercise programs for localized disaster response
- ☐ Create telecommunications regulatory staff emergency training program
- ☐ Implement joint exercises with emergency management agencies
- ☐ Develop senior leadership tabletop exercises for telecommunications failure
- ☐ Create exercise scenario library specific to telecommunications emergencies
- ☐ Implement after-action review process for telecommunications aspects of all disasters
- ☐ Develop lessons learned integration process for regulatory improvements

PUBLIC INFORMATION AND SUPPORT

- ☐ Create public education program on telecommunications during emergencies
- ☐ Establish standardized communications templates for service disruptions
- ☐ Develop backup communication options guidance for the public
- ☐ Create vulnerability mapping for telecommunications-dependent populations
- ☐ Implement emergency communications needs assessment for vulnerable communities
- ☐ Establish public information strategy for extended telecommunications outages
- ☐ Create community emergency communications center deployment strategy
- ☐ Develop accessible communications alternatives for people with disabilities
- ☐ Establish guidance for local governments on telecommunications restoration
- ☐ Create clear public messaging on service restoration priorities and timelines
- ☐ Implement social media strategy for telecommunications disruption updates
- ☐ Develop public-private partnership for emergency community connectivity

TECHNOLOGY AND CAPABILITY DEVELOPMENT

- ☐ Establish research program for emergency telecommunications technologies
- ☐ Create standardized interfaces for emergency communications systems
- ☐ Develop deployable integrated communications platforms for disaster areas
- ☐ Implement next-generation emergency alert system development
- ☐ Create national mesh network capability for emergency communications
- ☐ Establish alternative communications testing and certification program
- ☐ Develop high-altitude platform station (HAPS) capability for disaster coverage
- ☐ Create satellite communications rapid deployment capability
- ☐ Implement emergency telecommunications data analytics program
- ☐ Develop artificial intelligence applications for damage assessment
- ☐ Establish telecommunications resilience technology incentive program
- ☐ Create cybersecurity standards for emergency communications systems

Remember: As a government telecommunications authority, your role extends beyond regulation to enabling and coordinating the national telecommunications response to disasters. Your preparedness directly affects how quickly essential communications can be restored to communities, critical infrastructure, and emergency services during their time of greatest need.

MARCOS DUARTE RIVERO

Chapter Index

Chapter 4: Response Protocols and Procedures

Chapter 5: Training and Simulation Program

Chapter 6: Psychological Aspects of Emergency Response

Chapter 7: Technological Innovations in Emergency Response

Chapter 8: After the Storm: Recovery and Learning

MARCOS DUARTE RIVERO

This book is dedicated to and inspired by the tireless work performed daily by engineers, technicians, specialists, installers, and support staff throughout the telecommunications industry. Their unwavering dedication enables our constant connectivity in an increasingly interconnected world.

I extend my deepest gratitude especially to those anonymous heroes who work diligently to rapidly restore telecommunications services when disasters strike communities. These individuals brave adverse conditions, often placing themselves at risk, to reestablish the critical communication infrastructure that becomes even more vital during times of crisis. Their behind-the-scenes efforts ensure that emergency services remain functional, families stay connected, and communities begin the process of recovery.

Their technical expertise, combined with remarkable commitment to service, forms the backbone of our modern communication systems. Though their contributions frequently go unrecognized, the resilience of our telecommunications networks stands as a testament to their excellence and dedication.

To these professionals who maintain our connections through both ordinary days and extraordinary circumstances, I offer my most profound appreciation and respect.

Thank you.

Other books from BUSINESS FIRST AID SERIES

BUSINESS FIRST AID

"Saving companies"

https://www.businessfirstaid.online

www.ingramcontent.com/pod-product-compliance
Lightning Source LLC
LaVergne TN
LVHW090401210926
848480LV00002B/414
* 9 7 9 8 2 3 0 0 3 8 4 1 2 *